MARLBOROUGH

Cobb Valley
NW Nelson Mts
Mt Arthur
Boulder Lake & Brown Cow Pass

Nelson

Cook Strait

Lake Rotoiti
Travers Valley & Mt Travers

Molesworth
Woodside Gorge
Mt Benmore
Kaikoura Mts
Clarence R.

WEST COAST

Cupola Basin & Mt Cupola

Kaikoura

Lewis Pass
Mt Lucretia

Jollies Pass
Jack's Pass

Arthur's Pass
Castle Hill

Cass
Sugarloaf
Mt Oxford

Christchurch

Upper Rangitata
Mt Cook

Torlesse Range
Fog Peak
Craigieburn Range
Porter's Pass

McKenzie Pass

Hollyford Valley
Homer Tunnel
Gertrude Saddle

Lake Hawea
Lake Wanaka

Mt Belle & Humboldt Mts
Old Man Range
Rock & Pillar Range

Lake Wakatipu

Wapiti Lake

FIORDLAND

inton Valley

Eyre Mts

Dunedin

Mt Maungatua
Garvie Mts

Foveaux Strait

L & S licence 1984/27

A Field Guide to the
Alpine Plants of New Zealand

A Field Guide to the

ALPINE PLANTS

of New Zealand

John T. Salmon

GODWIT

Published by Godwit Press Limited,
P. O. Box 4325, Auckland 1,
New Zealand

First published 1968
Second edition 1985
Third edition 1992
Reprinted 1993
Printed in Singapore

National Library of New Zealand
Cataloguing-in-Publication data

Salmon, J. T. (John Tenison), 1910–
Field guide to the alpine plants of New Zealand/
John T. Salmon. 3rd ed. Auckland, N.Z.: Godwit Press, 1992.
1 v.
First ed. published: Wellington, N.Z.: Reed, 1968.
Includes bibliographical references and index.
ISBN 0 908877-17-X (hbk.)
1. Alpine flora—New Zealand. I. Title.
II. Title: Alpine plants of New Zealand.
581.90943

CONTENTS

To my sons Guy, Ian, and Roy,
who came with me on many happy excursions
into the alpine areas of New Zealand
during the preparation of this book

PREFACE

During many years working as an entomologist, I became aware of the need for a well-illustrated field guide to the alpine plants of New Zealand. This need was confirmed by the many people I met who enjoyed the shrubs and flowers found in abundance in our mountain regions and wished to identify them. In 1968 my first *Field Guide to the Alpine Plants of New Zealand* was published.

This new book is a complete revision of the original field guide and has been corrected and updated in accordance with changes in the taxonomy of New Zealand plants. It is enhanced by the addition of further species. It is not intended as an exhaustive survey but rather as a guide to plants likely to be encountered in the mountains. Common plants are included as well as those less frequently seen because they occur only locally or are small, specialised or not easily found.

Although each plant is dealt with individually, it should be remembered that plants usually grow together in varied, complicated and often extremely lush associations. In fact, the more closely intermingled many of our alpine plants become, the better they seem to grow. It is these lush associations and their adaptations to growing in a very specialised alpine environment that make New Zealand's mountain vegetation so fascinating and so beautiful. The extent and nature of some of these associations are illustrated in the concluding section of this book.

A plant is usually noticed in a particular place or environment and, for this reason, my pictures are arranged ecologically in the zonal order in which one is likely to encounter them. I believe such an arrangement is more useful, especially to the novice, as well as being most interesting. Superimposed upon this, where possible, are groupings of plants belonging to large genera. For example, the genus *Aciphylla* is most conveniently brought together in the grassland section where these plants predominantly belong. Each photograph has been designed to show the essential features of the plant it illustrates so that positive identifications may be made. With most of us the ability to name a plant when we see it helps us to remember it and to recognise it again in new situations. Identification, therefore, becomes an important and valuable aid, which widens our knowledge of the natural history of plants, their habitats and their ecology.

The correct naming of a plant often depends upon a close examination of its flowers, leaves or stems. This is particularly so with shrubby alpine plants, and to illustrate many species I have used close-up and high-magnification photography in preference to photographs of the whole plant, which would be of little real value. To assist in identification, the picture captions supply additional information, such as relative sizes and measurements, not shown in the photographs. The locality where the plant was found, together with the month during which the photograph was taken, is recorded with each picture and, to assist in further study and in the identification of closely related species, the family in which the plant is classified in botanical literature is also included, either individually for each picture, or collectively for large genera such as *Hebe, Aciphylla, Dracophyllum.*

While botanical nomenclature serves the specialist admirably, scientific names as such are enigmas with little meaning to the ordinary person. For this reason, every plant illustrated in this book has a common or popular name recorded along with its scientific name. The Maori names for shrubs and trees are well known to most New Zealanders. These names are generally pleasant sounding and meaningful, but the Maoris had named very few of our alpine plants, which made the selection of popular names for this book a little difficult. Accordingly, I have used names generally accepted where these exist, and where none exist I have coined new names. I make no excuse for this as I firmly believe that the free use of popular names will help enormously in encouraging amongst New Zealanders a greater appreciation of our alpine plants.

Many friends and colleagues have helped me in the preparation of this book and to all of them I express my sincere thanks. I would especially mention Professor W.R. Philipson, previously of Canterbury University; Dr Graeme Ramsay, Entomology Division, DSIR, Auckland; Mrs Margaret Parsons, Benmore P.B., Blenheim; Mr A.H.C. Christie, Animal Ecology Section, N.Z. Forest Service; Mr John Henderson of the N.Z. Deerstalkers Association; Mr John Anderson of Albury; and Miss Nancy Adams of the National Museum for assistance in locating specimen plants to photograph. I am grateful to Dr J.W. Dawson, Mr Bruce Sampson and Mr B. Sneddon of the Botany Department, Victoria University, for assistance in the identification of specimens. To Mr A.P. Druce, Botany Division of the Soil Bureau, Lower Hutt, I owe special thanks for much valuable assistance in checking identifications and nomenclature, and making numerous suggestions in the preparation of the text.

And lastly but by no means least, my thanks to my wife, Pam, for her help and patience, not only in the field, but especially in keeping a watchful eye on my writing and checking the text of this book as she has done so effectively with all my other books.

J.T. Salmon,
Taupo, April 1984.

ECOLOGICAL DIVISONS

The plates are arranged in six ecological divisions corresponding to the recognisable zones of subalpine and alpine scrub; tussock grasslands and carpet grasslands; screes, rocky and stony places; bogs and swamps; streamsides, damp or shaded places; herbfields and fellfields.

A seventh section deals exclusively with the alpine plant associations.

1. **Subalpine and alpine scrub** is a zone of small trees, shrubs, and herbaceous plants forming a transition from the forest to the tussock lands, grasslands, or fellfields of the alpine zone.

2. **Tussock grasslands and carpet grasslands**, often called **meadows**, cover very extensive areas in the mountain regions of New Zealand. Herbaceous and other plants infiltrate into them, but these areas always remain dominated by tussocks and grasses.

3. **Screes, rocky and stony places** are, for the most part, bare stony areas populated sparsely by very special forms of plants highly adapted for life in these exposed, often arid regions over which great extremes in climate prevail.

4. **Bogs and swamps** occur throughout our mountains as areas in which free water can appear on the surface. They may be extensive, as at Key Summit, or small and local, perhaps only a square metre in extent, forming around seepages and in hollows amongst scrub, tussock lands, herbfields and fellfields.

5. **Streamsides, damp or shaded places** do not normally produce free water at the surface but remain sufficiently wet or shaded to maintain plants requiring moist root runs. Such places occur throughout scrub, tussock lands, herbfields and fellfields, in the lee of banks, near seepages and in the shade of shrubs.

6. **Herbfields and fellfields** are found mostly on the higher alpine slopes, in cirques and valley heads. They are often characterised by luxuriant growth and may develop from or in tussock lands or amongst the relatively stable rocky zones of the mountains.

PLANTS OF ALPINE AND SUBALPINE REGIONS

The mountains of New Zealand are nearly always clothed on their lower slopes by heavy forests dominated by either mixed podocarp species or beech (*Nothofagus*). These extend up the spurs to approximately 770m in the far south and to 1,050-1,250m in the north and must be traversed by anyone who wishes to reach the alpine regions above. The average height of our mountain ranges lies between 1,550m and 1,850m and, lying as New Zealand does within the temperate climatic zone, the alpine botanic regions occupy roughly the altitudinal range between 900m and 2,000m, which makes them, therefore, relatively easy of access.

The highest mountains in New Zealand are in the South Island, and amongst these occur the richest section of the alpine flora and the most exciting and beautiful associations of New Zealand plants. Almost half the plant species found in New Zealand grow amongst the high mountains and never normally descend to lowland levels. The lush growth and richness of our alpine plant associations in their primordial condition are features of our New Zealand mountains that never fail to draw surprise, interest and admiration from all who take the trouble to visit them.

The importance of preserving unharmed this luxuriant natural cover, which acts as a gigantic sponge during heavy rain, cannot be overemphasised. Though vegetation has suffered great damage over the last few decades from the depredations of introduced browsing animals, the culling activities of government hunters combined with a stricter control of sheep grazing has, in a number of areas, allowed a resurgence of growth. This regeneration is heartening to see and in some places, such as the Boulder Lake region, northwest Nelson, it has been quite remarkable.

When discussing the alpine plants of New Zealand, it is usual to designate the upper or alpine belt as that region above the normal winter snowline. The subalpine belt is that region lying below this line but above the lowest temporary winter snowline, down to which snow may persist for two to three weeks at a time. Below this belt snow may descend to lower levels during severe storms and persist there for several days, giving definition to a further lower subalpine or montane belt of vegetation. The altitudinal range of these belts varies from place to place, but, in general, it is lower in the south than in the north.

Each of these alpine belts is characterised by certain very distinct kinds of plants. Each different mountain locality, and in fact each mountain, forms an ecological zone peculiar to itself. The nature and extent of each ecological zone depends upon a number of factors, among which the more important are the direction towards which the general slope faces—whether it is east (dry) or west (wet), north (warm) or south (cold); the nature and extent of the rainfall; the exposure to strong winds; the angle or steepness of the slopes; and the nature of the underlying soils and rocks. Plants found in alpine and subalpine regions may be growing in scrub, tussock grasslands or carpet grasslands, meadows or herbfields, exposed rocky or stony places, stony places in scrub, grasslands and herbfields, and in fellfields, all of which may be either damp or dry. In addition to these situations there are the mountain tarns, bogs and swamps, which together form extensive areas inhabited by many kinds of water-loving plants, the shady banks alongside streams and waterfalls, and the low, damp, shaded banks found in steep, rolling country and around rocky outcrops. All these types of habitats that occur time and time again in infinite variety all over the mountains provide situations that different kinds of plants find favourable to their development and growth.

As one ascends through forest to higher slopes, a progressive change in the vegetation becomes noticeable after 600m has been reached. Previously dominant species give way to others, and as the region of the 770-900m zone is entered, marked changes become apparent. The trees usually begin to diminish in size and to thin out, while montane and subalpine species of shrubs such as pepper trees (*Pseudowintera*) and species of *Coprosma, Hebe* or *Olearia* increase in number and frequency of occurrence. Towards the upper limit of the forest the trees become smaller and are often stunted. They finally peter out as one enters the zone of the subalpine scrub. Often there is a mingling of these two belts of vegetation with one another and no hard-and-fast line of demarcation can be drawn.

However, above the forest line a clear edge to the forest is normally discernible, and in some South Island districts, where pure beech forest predominates, the transition from forest to subalpine scrub or tussock lands may be much more abrupt. In some parts of Fiordland stunted beech trees take the place of subalpine scrub and the transition may be directly from forest to alpine meadows, tussock lands or herbfields. This type of transition is clearly seen on Key Summit, where small, perfectly formed but old beech trees, sometimes only 2-3m high, grow as a patchy scrub or as isolated specimens. In the beech forest of northwest Nelson, also, transition from the forest to tussock lands or meadows is abrupt. The screes that reach down into these forests are abnormal, resulting from human interference with the natural vegetative balance of the regions. In the vicinity of the Lewis Pass and Arthur's Pass the beech forests pass abruptly into definite belts of a subalpine scrub-herbfield type of vegetation. All this demonstrates the difficulty of clearly defining in the field any theoretical zonation of alpine vegetation.

However, in spite of these difficulties, zones of subalpine scrub and, at higher elevation, zones of alpine scrub, tussock grasslands, herbfields and fellfields can be recognised. The disposition of these zones depends largely on climatic factors and the conditions of the terrain. The annual total precipitation of snow, rain and mist has a far-reaching effect, largely through the incidence of erosion, upon the species of plants present and the closeness in which they grow in any alpine association. Fellfields are a rather open type of plant association consisting primarily of low-growing species able to establish themselves and grow in unstable, rocky and stony places, which may be on either the wet or the dry side of the mountains.

The instability of these areas arises from the friable nature of the rocks forming the substratum, and this leads to a high rate of erosion. On an unstable substratum close, dense associations of plants never have time enough to develop to any great extent. This is because a fellfield type of association is continually being destroyed and renewed in the ebb and flow of the struggle between plant life endeavouring to establish itself and these destructive though natural forces of erosion. As erosion slows down and slopes become more stable, plant life establishes itself more easily. Associations grow denser and new environmental conditions arise that enable new species to enter and establish themselves until, finally, a relatively stable and luxuriant zone of vegetation can be recognised; this is known as a herbfield. Herbfields are characterised by an abundance of large herbaceous plants, and true herbfields consist almost entirely of herbs and low shrubs without any, or at least with very few, tussocks. Shrub herbfields are dominated by shrubs and the herbaceous plants occupy the spaces between the shrubs. Similarly, a tussock herbfield is dominated by tussocks with herbaceous plants between.

On many mountain slopes of this kind tussock and carpet grasses have become the dominant plants; these areas are the mountain tussock grasslands and carpet grasslands so familiar to trampers and climbers.

Almost anywhere in fellfields, herbfields, tussock, grasslands or scrub, seepages from springs can arise and create damp or wet and boggy areas that are invaded by plants adapted to living in such places. Shady banks also attract special kinds of plants, which colonise these niches in the larger zones. Bogs and swamps can be very extensive, as on Key Summit, where occur some of the finest bogs of our mountain regions.

Tussock herbfields are often referred to as alpine meadows. These extend from the higher mountains of the Raukumara Ranges (on the East Cape of the North Island) southwards along the main mountain ranges of both islands into Stewart Island.

Alpine meadows do not always exhibit closely growing, dense plant associations. In certain areas, on account of climate and other conditions, they are sparse and open. This, however, is rare and normally the vegetation of alpine meadows in their unspoiled condition forms a close, dense association of tussocks and herbs, so dense, in fact, that they can be difficult to walk through. Conditions like this can be encountered at the present time in certain areas of the Tararua Ranges, on the mountains around Boulder Lake, in Cupola Basin, on Mt Lucretia above the Lewis Pass, in Arthur's Pass and in Fiordland. It is in such places as these that the true wealth of New Zealand alpine plants can be seen and appreciated.

The principal plants that enrich these alpine associations are species of buttercup (*Ranunculus*) flowering in the spring, followed by mountain daisies (*Celmisia*), the Maori onion (*Bulbinella*), bluebells (*Wahlenbergia*), *Ourisia* species and the mountain eyebrights (*Euphrasia* spp.) throughout the summer, and the gentians in the autumn. These present a wealth of flowers over a long period, though perhaps it is a little unfortunate that white and yellow colours predominate amongst the flowers of our alpine regions to the almost complete exclusion of red and blue.

Alpine and subalpine scrub. Scrub in the mountains contains a wealth of plant species ranging from creeping, mat-forming plants, through herbs, to large shrubs of *Olearia, Brachyglottis, Dracophyllum, Coprosma* and *Hebe.* Species of *Hymenanthera, Aristotelia, Gaultheria, Halocarpus, Lepidothamnus, Cassinia, Pseudowintera, Pittosporum, Pseudopanax* and *Coriaria* all contribute to the plant communities that make subalpine scrub associations interesting and often very beautiful. Mountain flax is as common a plant in subalpine scrub as it is in herbfields and rocky places. Among the *Dracophyllum* species *D. traversii* is the largest. It is a fairly common plant throughout northwest Nelson and forms small forests along the leading ridge from the Flora Saddle to Mount Arthur, and on the Canaan Track to Moor Park in Abel Tasman National Park. In these grass-tree forests the long, thin leaves of the grass tree persist for a long time on the ground without decaying, producing a rich brown carpet beneath the trees.

Brachyglottis rotundifolia var. *elaeagnifolia* is a handsome shrub that occurs in the North Island scrub. An almost pure stand and one of the best I have seen of this shrub grows just below the summit of Mt Tauhara, near Taupo. Tupare (*Olearia colensoi*), another fine North Island species, grows in extensive stands on the Ruahine Ranges. One of the most accessible of these almost surrounds Wharite Peak, where a television translator was erected. The variation that can occur between the North Island and the South Island plants of one species is well shown by hakeke (*Olearia ilicifolia*), which is narrow-leaved on the Tararuas but broad-leaved in the South Island mountains.

Subalpine scrub can become extremely dense and luxuriant when unharmed by animals and in this condition is best suited to perform its protective function of water conservation and erosion control. It merges almost imperceptibly into tussock grassland or herbfield, and many herbaceous plants of the mountains grow equally well in either scrub or tussock herbfield.

Alpine and subalpine grasslands. Large areas in the New Zealand mountains are clothed with tussock and carpet grasses, forming meadows that extend in some places to heights of 2,000m or more. Tussocks dominate these native grasslands and good specimens of red tussock (*Chianochloa rubra*), mountain tussock (*C. pallens*), or hunangamoho (*C. conspicua*) may be 2m high. Between these tussocks in a healthy, stable meadow the ground is covered by grasses, sedges, creeping mat-plants and herbs such as *Pratia, Ranunculus, Acaena,* bluebells and orchids. There are many species of tussock that belong to the genera *Poa* and *Festuca* as well as *Chianochloa. Rytidosperma viride* (silver tussock) is another widespread tussock found throughout our mountain regions.

Wild spaniards or speargrasses are common plants often met with in grasslands, herbfields or scrub. With their stiff, pungent, sword-like leaves, an encounter with one of the larger species of these plants can be very unpleasant. Some of the smaller speargrasses, however, such as *A. monroi* or *A. congesta,* have soft leaves and are quite harmless to touch. The *Acaena* species are common in grasslands and *A. novae-zelandiae,* the bidibidi or piripiri, is noted for its seed heads, which cling to clothing and animal fur. Tussock grasslands merge with herbfields, and tussock herbfields and herbfields cover vast areas throughout the mountains of New Zealand.

The giant buttercup (*Ranunculus lyallii*) is best sought in flower, during December or early January, in the herbfields near Mt Cook, Arthur's Pass, the Harris Saddle, the Homer and Gertrude Cirques at the head of the Hollyford Valley, and on the MacKinnon Pass. At the Homer I have seen acres of this plant in flower during December, an unforgettable sight of glistening nodding panicles in the early morning sun.

Korikori, the yellow mountain buttercup (*Ranunculus insignis)*, though not as large a plant as *R. lyallii*, is, nevertheless, a most spectacular and beautiful plant. There is a fine colony at about 1,550m on Mt Ruapehu in the valley west of the Pinnacles. These plants usually flower during early January. Splendid examples can also be seen in the rocky funnels on Mt Holdsworth at about 1,250m, flowering during December. In the herbfields on the higher slopes of Mt Holdsworth *R. verticillatus* is common and is in flower at the same time.

The mountain foxgloves (*Ourisia* spp.) are usually shade-loving and *O. macrophylla* makes a wonderful display on Mt Egmont during January, when its flowers spread over the shaded banks of the tracks leading from Dawson Falls towards the Stratford Plateau. *O. macrophylla* var. *lactea* occurs quite commonly among the tussocks on Mt Lucretia and at Cupola Basin, the plants sharing their shade with a prostrate form of *Ranunculus multiscapeus*. Above Temple Basin *O. macrocarpa* also grows abundantly in the shade of high tussocks.

The *Brachyglottis* spp. are among the most showy of our alpine plants. The little *B. bellidioides* always puts on a brilliant display at the height of its flowering season, which is about mid-January at Arthur's Pass.

The yellow and white snow marguerites (*Dolichoglottis lyallii* and *D. scorzoneroides*), when alongside streams or seepages, quite often grow into large clumps or colonies that produce magnificent displays of flowers during summer. The white marguerite is a very strong, erect plant, which becomes smothered with blooms and can be seen in quantity at its best during early and mid-January along the banks of the little streams in Temple Basin and elsewhere at Arthur's Pass, or in the Gertrude Cirque at the Homer.

The mountain daisies (*Celmisia* spp.) are by far the most plentiful of our alpine plants, with many diverse forms adapted to almost all types of mountain habitat and always flowering in profusion during summer. These gay plants are everywhere in our mountain regions, forming large masses in herbfields, reaching out from among tussocks, brilliant amongst the sombre scrub or sprouting from rocky crevices, their flowers dazzling in the summer sun or welcoming like small beacons in the cold blanket of the mountain mist.

Everywhere we go in our mountains during spring and summer the New Zealand bluebells hold up their delicate flowers towards the sun. These vary in colour from pale blue to white; the fresh flower is usually a darker blue, fading as it ages, but the intensity of the colour does vary from locality to locality and appears to be affected by the type of soil in which the bluebell is growing.

February is the time when the eyebrights bloom, and during this month *E. cuneata* and *E. drucei* produce cascades of flowers all along the Ruahine and Tararua Ranges. Peeping out from under tussocks, overhanging banks, and amongst rocks, these lovely flowers keep the mountains bright until the gentians take over and continue the floral display into April or even May. Many species of gentians are found all over the New Zealand mountains, and on the slopes of Mt Hector in the Tararuas I have seen thousands of the alpine gentian (*G. patula*) in bloom together during March. Gentians are also common amongst the mountains of northwest Nelson, where they flower in great abundance during late summer on the alpine meadows of Abel Tasman National Park, in Cupola Basin and on the slopes above Boulder Lake.

Plants of bogs and swamps, streamsides, damp and shady places. Bogs and swamps, and damp and shady places are always worthy of investigation, for here grow many fascinating plants. The insectivorous plants (*Drosera* and *Utricularia* spp.) thrive in bogs, and *D. spathulata* and *D. binata* are two common and most spectacular species in sphagnum bogs. *Utricularia monanthos*, the purple bladderwort, bears an exquisite little flower during January (fig. 272). This plant has some of its leaves changed into hooded underwater traps in which water insects inhabiting the swamps are caught for digestion by the plant. In January 1967 I saw this bladderwort flowering in thousands all through the bogs of Key Summit; it was an unforgettable sight. In many previous visits to Key Summit I had never seen the flowers of this little plant. The 1967 season in Fiordland had been exceptionally fine with no rain, and the bogs had dried up to the extent that small ponds were empty. Perhaps this bladderwort flowers only under these semi-dry conditions.

The extensive bogs on Key Summit contain a wealth of interesting plants ranging from sundews, bladderworts, orchids, and the little lily *Herpolirion novae-zelandiae* to the bog forstera, bog daisy (*Celmisia alpina*) and larger species such as *Dracophyllum pronum* and bog pine (*Halocarpus bidwillii*). These sphagnum bogs of Key Summit are typical alpine bogs of high rainfall areas. Sphagnum moss is able to absorb and store very large quantities of water. It usually grows in depressions and basins where the streams cannot remove the water as fast as it accumulates. As the sphagnum mound grows upwards, its lower part dies and forms peat, on which many bog plants thrive and grow in association with the sphagnum. The hard, compact mounds of the alpine cushion (*Donatia novae-zelandiae*) are common everywhere near seepages and in bogs. Readily accessible cushions of this plant occur in the bogs near the summit of Arthur's Pass, where they flower profusely during January. Above about 1,500m, amongst the rocky outcrops of mountainsides and tops, small seepages often produce damp spots. Here the rock cushion (*Phyllacne colensoi*), which is very similar in appearance to *Donatia*, the alpine cushion, fits tightly into the spaces between the stones or spreads out to cover them with a bright yellow-green mound that springs to life during January to glow with creamy-white flowers. Other plants that can nearly always be located in damp areas belong to *Gunnera, Cotula, Nertera* and *Forstera* genera, while in shaded places panakenake (*Pratia angulata*) can always be depended on for fine displays of blooms throughout the summer.

Gunnera plants are common in shaded wet places. The finest displays of these plants I have seen occur in the upper Travers Valley where, in places, *G. dentata* extends along the riverbanks as patches 6-10m long, producing a wonderful splash of orange-red during April and May. *Gunnera* plants forming patches like this also occur in the Mt Cook region, where they are usually bright with berries during April. On Mt Egmont, around Wilkie's Pools, the shaded banks, wet from seepages, provide ideal conditions for the leafy forstera, which thrives and flowers here with great abandon.

Small mountain streams splashing over stones amongst tussock lands and herbfields keep their banks moist, and in these places *Ranunculus gracilipes, Parahebe* spp. and other water-tolerant plants will often thrive. *Parahebe catarractae* and *P. lyallii* plants growing in the cool spray of a waterfall usually form large, lush, luxuriant bushes. I have seen these species growing like this near Boulder Lake and along the banks of the Cupola River.

Plants of screes, rocky and stony places. The mountains of New Zealand have always been subject at the higher altitudes to wind and water erosion, which together cause the flaking of rocky outcrops that produces shingle slides or screes. Although these phenomena have been increased to alarming proportions over the last 100 years by the impact of European man and his browsing animals, the alpine scree has been a natural habitat for special plants for a very long time. Plants adapted to growing on screes are deep rooted, drawing their water from the moist ground that lies below the moving cover of stones. As with all plants that grow in rocky places, scree plants can withstand extremely high day temperatures when their rocky surroundings are heated up by sunshine, and very low temperatures at night because of the rapid loss of stored heat from the surrounding rocks and stones by radiation.

The places in which they live are always steep, exposed and very windy, and the winds that blow across them are often very cold. Dust arising from the eroding screes is often blown in clouds across these regions, and plants living there must be able to withstand the effects of wind-blown dust and sand. Scree plants are, therefore, mostly low growing, creeping and succulent; many of them are grey in colour, matching the stones amongst which they live, and they are therefore often very difficult to find. However, it is remarkable how many of these plants, once seen, can be located on a small area of a single scree.

Scree plants are often overwhelmed by moving stones but their stems grow outwards again towards the light. These plants have a perpetual struggle for survival amongst the cascades of stones continually descending upon them. This phenomenon can be seen with Haast's buttercup, the scree lignocarpa, *L. carnosula,* and *Stellaria roughii.* Long fine stems of Haast's buttercup can sometimes be followed down deep through the stones to a stout rhizome from which fine roots penetrate far into the cool, damp substrata of fine rocks and sand.

On screes specially adapted plants with long, wiry stems creep among the stones, binding them together and helping to stabilise them. *Cotula atrata, Cotula pyrethrifolia* and *Acaena glabra* are plants of this type found only on screes. Creeping pohuehue (*Muehlenbeckia axillaris*), creeping matipo (*Myrsine nummularia*) and mountain cotula (*C. pyrethrifolia*) often occur along the edges and bottoms of screes, creeping over the stones and binding them together. *Leucopogon suaveolens,* mountain totara (*Podocarpus nivalis*), *Coprosma pseudocuneata* and similar plants also play their part in populating and stabilising rocky outcrops and screes.

Perhaps the two most unusual scree plants are the penwiper plant (*Notothlaspi rosulatum*) and the mountain cress (*Notothlaspi australe*). The latter is quite common in Cupola Basin, where I have seen many plants on screes facing both north and west. When in flower, mountain cress pours out a glorious sweet scent, which floats away from the plant in waves. This perfume is so strong that I have detected the presence of flowering plants before I have seen them. The penwiper plant is more difficult to find, as it appears to be an itinerant plant, widely distributed but of irregular occurrence in any given region. It should be looked for among the smaller stones that are becoming stabilised along the edges of screes above 1,300m, where the substratum forms into a moist soil. The flowers of the penwiper, like those of mountain cress, produce a very strong, sweet scent, and even tiny plants not more than 3 cm across may bear a few flowers.

Plants that grow in rocky places such as cirques, cliff faces and rocky outcrops must, like scree plants, be able to withstand great extremes of temperature between daytime and night-time, summer and winter. The places in which they grow are nearly always exposed and subject to high-velocity winds. Frequently they grow in wind funnels between crags where, during winter, snow may lie feet deep. They obtain their water from deep-running roots, which penetrate the cracks that criss-cross the rocks, and when the snow melts in glaciated cirques, bands of vegetation flourish along these cracks as well as along the numerous small water-courses.

Plants of the genus *Raoulia* have evolved and adapted themselves in a wonderful fashion to living among rocks and stones. On stony riverbeds are the scabweeds such as *Raoulia australis, R. haastii* and *R. hookeri,* while on the heights the vegetable sheep *R. eximia, R. mammillaris* and *Haastia pulvinaris* are quite at home under extreme conditions of climatic variation. Vegetable sheep belong to the daisy family and are among the world's most extraordinary plants. They form huge, thick, tight hummocks that sprawl over stones and rocks. They are extremely hardy, and resistant to wind and falling stones; they can be buried by winter snows; they can even be walked upon without suffering damage.

In high, sheer, rocky places of the South Island *Helichrysum intermedium* grows out of cracks and crannies in the rocks, and on the screes below, *Leucopogon suaveolens* and mountain totara help to bind the sliding debris. Mountain hebes are common in more gently sloping rocky places away from screes, and cover large areas in almost pure stands, such as occurs with *Hebe odora* in the upper Hollyford Valley.

No book of alpine plants would be complete without mention of the edelweiss, a plant whose name is synonymous with high mountains. In New Zealand we have two plants called edelweiss — *Leucogenes grandiceps* being the South Island plant, and *L. leontopodium* the North Island edelweiss. The South Island edelweiss is found in stony and rocky places and amongst rocks in herbfields. The North Island edelweiss grows abundantly in rocky places amongst herbfields and is particularly common on Mt Holdsworth in the Tararuas, where it flowers in great profusion during January and February.

Ecology. From studies of the associations of plants in the mountains, one learns, among other things, of the relationships that exist between one species and another, and between them collectively and their physical environment; of the complexity of plant response to the limiting factors of sunlight, climate, soil, temperature and water; of the intimate associations that have evolved between plants themselves and between them and the soil or rocks upon which they grow; and of the vital importance of an adequate plant cover in mountain regions for water conservation, erosion control and soil formation.

The web of plant life is a wonderful phenomenon. It reveals itself in closely knit but well-established and oft-repeated associations of certain species; in zones or belts of vegetation of certain types linked to altitude and to climate; or in the development and growth of specialised forms adapted to particular conditions or particular purposes.

The plant associations found in New Zealand mountains are many and varied and of great interest as such to the plant ecologist. Today many of these associations are undergoing rapid changes brought about through the impact of browsing animals. To ensure the conservation of our alpine vegetation, ecologists must find solutions to the problems these animals are creating.

The native alpine flora of New Zealand is unique and its preservation in our own immediate interests as well as for those of posterity is a duty that falls heavily on each and every New Zealander. The vegetation with which Nature clothed our mountains took millions of years to evolve and is the most perfect protective cover possible. Through the impact of animals and the practice of periodic burning of high-country tussocks, the composition of alpine plant associations has been drastically altered in many places; in some places these associations have been virtually destroyed, and accelerated erosion on a vast scale has set in. This is the case in the Craigieburn Mountains and on sections of the Torlesse Range and other ranges of Canterbury, where the erosion cycle is so far advanced that reclamation of the heights may prove impossible to accomplish.

Deterioration of alpine meadows leading to accelerated erosion can be seen almost anywhere amongst the mountains of New Zealand and, in the interests of our agricultural economy, this process must be halted or at least slowed down. A proper appreciation of our alpine plants and a nationwide understanding of their vital role in ecology is a first prerequisite to the control of these destructive processes of erosion.

SUBALPINE AND ALPINE SCRUB

1. Tupare plant in flower, Wharite Peak (December)

Tupare, leatherwood *Olearia colensoi*

Tupare forms a closely branched shrub up to 3m high, with the branches clothed in fawn-coloured wool. The thick, leathery, serrated leaves, 8-20cm long, are clad below with dense white or brownish-coloured wool. The flowers, 2-3cm across, occur from November to January, and their seeds appear from December till February. Tupare is found throughout New Zealand in subalpine scrub from Mt Hikurangi southwards, coming down to sea level in Stewart Island. It grows either as isolated specimens, sometimes in tussock meadows, or in dense local stands as, for instance, here on Wharite Peak at the southern end of the Ruahine Range.

Family: Compositae

2. Hakeke flowers, South Island form, Hollyford Valley (January)

3. Hakeke flowers, North Island form, Ruahine Mountains (January)

4. Spray of leaves of rough-leaved tree daisy, Moorpark, Abel Tasman National Park (November)

Rough-leaved tree daisy *Olearia lacunosa*

A much-branched shrub or small tree, up to 5m high, with its branchlets, leaf petioles, leaf undersides and flower stalks all clothed in a thick, brownish or rust-coloured wool.

The leaves, up to 17cm long and 2-5cm wide, have very prominent midrib and veins, which impart an alveolate appearance to their lower surfaces as shown in figure 4. Found in subalpine and alpine scrub between 900 and 2,300m from the Tararua Ranges southwards. Fine specimens of *O. lacunosa* grow round the sink holes along the main ridge of Mt Arthur, behind Nelson, and along streamsides in the Moor Park area of Abel Tasman National Park.

Family: Compositae

Hakeke *Olearia ilicifolia*

A musk-scented shrub or small tree, up to 5m high, found in subalpine forest and scrub from the Bay of Plenty southwards. The strongly dentate leaves of most North Island plants are relatively narrower than those of South Island plants. Flowers occur in large corymbs from November to January. Hybrids between the North and South Island forms can occur. Hakeke often grows luxuriantly around clearings, along stream banks and along openings made by tracks. Quite large specimens grow in the upper Hollyford Valley.

Family: Compositae

5. Flowers of hard-leaved tree daisy, *O. cymbifolia,*
   Mt Terako (January)

6. Hard-leaved tree daisy *O. nummularifolia* with flower buds,
   Gertrude Cirque, Homer (January)

7. Flower spray of musk tree daisy, Gertrude Cirque (January)

Musk tree daisy   *Olearia moschata*
A much-branched tree daisy up to 4m high, producing a strong, musk-like scent, especially during hot weather. The close-set leaves are 8-15mm long and up to 5mm wide, usually glabrous above but with an appressed, white tomentum below. The flowers occur from November to March, and the plant is found in subalpine scrub up to 1,500m altitude from the Lewis Pass southwards. The form illustrated here may be a hybrid.

Family: Compositae

Hard-leaved tree daisies   *Olearia nummularifolia* and *Olearia cymbifolia*
Shrubs up to 3m high with branchlets that are sticky in the young stage and clothed with white or yellowish, star-shaped hairs. The closely set leaves on short 1mm long petioles are very thick and leathery; in *O. nummularifolia*, they are 5-10mm long by 4-6mm wide, with recurved margins and white or yellowish tomentum below. In *O. cymbifolia*, leaves are 6-14mm long, with the margins revolute almost to the midrib, giving a rather swollen appearance to each leaf, and young leaves are very sticky. Flowers about 5mm across occur from November to April in *O. nummularifolia* but in *O. cymbifolia* seldom occur after January. *O. cymbifolia* is found throughout the South Island only, whereas *O. nummularifolia* is found in both islands from the Volcanic Plateau southwards. Both species occur in subalpine and alpine scrub.

Family: Compositae

8. Flowers and foliage of Nelson mountain groundsel, Marlborough Mountains (December)

Nelson mountain groundsel   *Brachyglottis laxifolia*

This loosely branched shrub with its grey-green leaves reaches a height of 1m and, from December to February, smothers itself with these lovely bright-yellow flowers, each 2cm across. The branchlets and the undersides of the leaves are clothed with dense, short, white hairs, and the leaves are up to 6cm long by 2cm wide. The plant is found among the mountains of Nelson and Marlborough between 900 and 1,500m.

Family: Compositae

9. Thick-leaved shrub groundsel showing new leaf growth, Mt Holdsworth (February)

Shrub groundsel   *Brachyglottis elaeagnifolia*
This shrub, which grows to a height of 3m, is easily recognised by its
grooved branches and the pale to dark-brown, appressed hairs that clothe
the branchlets, petioles, flowers, stalks and lower leaf surfaces. The leaves
are 10-12.5cm long and 7.5cm across, with conspicuous veins. The flowers,
8mm across, occur in panicles up to 15cm long during January and
February. The shrub occurs commonly in subalpine scrub, particularly
along exposed margins and in the adjacent subalpine forest. It is found
in the North Island from East Cape southwards. A rather similar but
smaller shrub, up to 1.5m high, with smaller, more ovoid-shaped leaves,
2-2.5cm long by 1-2cm wide, occurs in subalpine scrub on Mt Holdsworth.
This species, *B. adamsii*, is readily recognised by the very sticky nature
of its branchlets, leaves and flowers, and its prominent midrib to the leaf.
                                                       Family: Compositae

10. Shrub groundsel with flowers, Mt Holdsworth (January)

Thick-leaved shrub groundsel   *Brachyglottis bidwillii*
A tightly branched shrub reaching 1m in height, with thick, leathery, elliptic
or obovate leaves 2-2.5cm long. The branchlets and undersides of the leaves
are densely clothed with soft, white to fawn-coloured hairs. The flowers,
1.5cm across, appear during January and February, and the shrub is found
in mountain scrub and fellfields from Mt Hikurangi southwards.
                                                       Family: Compositae

11. Flowers of the Kaikoura shrub groundsel, Mt Terako (December)

12. South Island shrub groundsel in flower, Homer
     Saddle (January)

Kaikoura shrub groundsel   *Brachyglottis monroi*
A spreading, much-branched shrub up to 1m high, with leaves 2-4cm long
and up to 15mm wide. Flowers 2cm across occur from December to March,
and the plant is found from 600 to 1,200m amongst the Kaikoura Ranges
and the Nelson mountains.

Family: Compositae

South Island shrub groundsel   *Brachyglottis revoluta*
A small shrub, up to 50cm high, with the branches spreading or decumbent
and the branchlets erect and densely leafy. Leaves are 3-6cm long and
2-3cm wide. The flowers are 2cm across and occur from January to March
at the top of ascending stalks, which may be 10cm long. Found in alpine
scrub and fellfields in western Otago and the Fiordland mountains.

Family: Compositae

Southern mountain shrubby groundsel   *Brachyglottis buchananii*
A compact, branched shrub, up to 3m high, with ribbed leaves on grooved
petioles 5cm long. The South Island counterpart of *B. elaeagnifolia,* which
it closely resembles. The leaves are 5-10cm long by 3-5cm wide, very glossy
above and with a whitish, rather thin wool below. Flowers 1cm across
occur in panicles up to 25cm long during January and February. Found
throughout the South Island in subalpine scrub and forest, mainly along
the west side of the mountains.

Family: Compositae

13. Southern mountain shrubby groundsel with flower buds, Arthur's
Pass (January)

14. Berries of mountain tutu.
    Volcanic Plateau (February)

15. Dense tutu with flowers,
    Arthur's Pass (December)

Mountain tutu   *Coriaria pteridioides*
A small shrub, reaching 60cm high, with square-shaped, slender, hairy branchlets; flowers and berries are borne on pendant stalks, the flowers from October to February and the berries from November to April. This tutu is found only in the North Island from the Volcanic Plateau to Mt Egmont, in subalpine scrub and grasslands.

Family: Coriariaceae

Dense tutu   *Coriaria angustissima*
A herbaceous plant forming large, dense patches and spreading by branching rhizomes, which bear erect stems up to 50cm tall. The narrow leaves are 7-10mm long, the flowers in racemes 3-5cm long, and these appear from November to February. This plant is found in subalpine scrub and rocky places in the South Island, more commonly on the western face of the mountains.

Family: Coriariaceae

16. Feathery tutu in flower, Mt Belle, Homer (December)

Feathery tutu   *Coriaria plumosa*
A small, bright-green shrub up to 40cm high, found from Mts Egmont and Hikurangi southwards to Stewart Island between 300 and 1,500m altitude, in grasslands, along streambeds and in scrub. The narrow, lanceolate to acuminate or acute leaves are 6-10cm long by 1.5-3mm wide, feathery in appearance on the stems. The tiny flowers that occur laterally on the four-sided branches appear from October till February and are followed by deep-crimson to black berries, which ripen from November to March. A very similar species, *C. pottsiana*, distinguished by having flowers terminal on the branchlets, is found on Mt Hikurangi.

Family: Coriariaceae

17. Mountain lacebark in flower, Lewis Pass
(January)

Mountain lacebark   *Hoheria lyallii* and *Hoheria glabrata*
Found in subalpine forest and scrub of the South Island, *H. glabrata*
mainly in the west and *H. lyallii* mainly in the east, these two deciduous
trees, 6-10m high, bear these delicate flowers, 2.5cm across, during January.
The delicate leaves, up to 6cm across, are thin and deeply lobed. *H. lyallii*
is common in subalpine scrub at Arthur's Pass and in the Hollyford Valley.
H. glabrata is common on the west side of the Lewis Pass, where frequent
copses of trees edge the road.

Family: Malvaceae

Common mingimingi   *Cyathodes juniperina*
This heath-like shrub with its small, pungent leaves, striped below as shown
here, common in lowland forest all over New Zealand, is also found not
infrequently in subalpine scrub. It grows to a height of 5m in lowlands
but seldom attains more than 2m at higher altitudes. The flowers, 3-4mm
long, normally occur from September till November and are followed from
January on by the white, pink, crimson or purple-coloured fruits 4-9mm
in diameter. In subalpine regions fruits of the previous season may remain
when flowers of the new season appear, as shown here from a plant at
1,000m on Mt Tongariro.

Family: Epacridaceae

Red mountain heath   *Archeria traversii* var. *australis*

A shrub, 2-5m high, with many erect or spreading branches. The leathery leaves are 7-12mm long, and the flowers, which arise in profusion from December to February, are 4-5mm long. This plant occurs in isolated patches in scrub and fellfields all through the mountains of the South Island up to an altitude of 1,250m.

Family: Epacridaceae

18. Red mountain heath in mass, Cleddau Valley (January)

19. Common mingimingi with both flowers and fruits, Volcanic Plateau (October)

20. Mountain totara showing fruits, Arthur's Pass (April)

21. Mountain totara with male catkins, Cleddau Valley (January)

Mountain totara   *Podocarpus nivalis*
A much-branching, prostrate, spreading shrub, 1-3m high, with thick-margined, leathery, closely set, spirally arranged and rigid leaves up to 15mm long. The slender male catkins (fig. 21) reach a length of 15mm, maturing during November and December, when they produce copious quantities of pollen. The female nuts, produced on separate plants, ripen from March to May, and when mature the peduncle or stalk becomes very much swollen and red-coloured, as shown in figure 20. Mountain totara occurs very commonly in subalpine and alpine scrub to an altitude of 1,600m from the Volcanic Plateau southwards.

Family: Podocarpaceae

22. Mountain toatoa with male catkins, Cupola Basin (December)

Mountain toatoa   *Phyllocladus aspleniifolius* var. *alpinus*
This is a shrub or small tree, reaching 9m in height, in which the leaf-like structures are not true leaves but modified branchlets called cladodes, which function as leaves. True leaves are produced only on young seedlings and soon give way to the cladodes, which are 2.5-6cm long, thick and leathery in texture, often with thickened edges. Flowers are produced from October to January, the male as catkins 6-8mm long, in clusters at the tips of the branches, the female as heads, 6mm in diameter when mature, singly along the stalks of the cladodes. Mountain toatoa occurs freely in subalpine forest and scrub from Hokianga southwards.

Family: Podocarpaceae

23. Mountain cottonwood with flower buds, Lewis
Pass (December)

Mountain cottonwood   *Cassinia vauvilliersii* var. *pallida*

Five species of *Cassinia* occur in New Zealand, but only the species *C. vauvilliersii* and its varieties are widespread in the higher alpine regions between 900 and 1,400m. Mountain cottonwood grows to 3m high and is recognised by its stout, grooved branchlets, usually covered by a sticky, brownish wool, and its narrow, leathery leaves, 5-12mm long, clad below by a brownish tomentum. Flowers occur from December till January as compact, sparkling corymbs, 25-40mm across, each with 10-20 capitula. The buds, especially of variety *pallida*, tinge with red when ready to open, and the display of seeds that follows in March and April adds to the seasonal attractiveness of this plant. Mountain cottonwood is very common in subalpine and alpine scrub and fellfields throughout New Zealand. The variety *pallida* is distinguished by having its branchlets and leaves clad with a greenish-coloured wool.

Family: Compositae

24. Flower head of Mountain cottonwood, Volcanic Plateau (January)

25. Mountain cottonwood in seed, Arthur's Pass (April)

26. Golden cottonwood with flowers, Upper Ure River Gorge (March)

Golden cottonwood *Cassinia fulvida* var. *montana*

A shrub growing to 2m high, with slender, sticky branchlets covered by a distinct, yellow-coloured tomentum that gives this plant a golden colour by which it can be easily recognised. It spreads readily, often covering large areas of hillsides with a warm, golden glow. The leaves arise each from an erect petiole and are 4-8mm long by 1mm wide, rather thick and leathery with slightly revolute margins. Golden cottonwood is found in subalpine situations from about Mt Hikurangi southwards, descending to lowland areas in the south. There are 35-40 capitula in each flower head, and each floret normally has a reddish edging. The plant is common at Arthur's Pass.

Family: Compositae

GRASS TREES

These are small to medium-size, generally erect shrubs or small trees belonging to the genus *Dracophyllum*. Their needle- or sword-like leaves arise erect from the stems. There are 35 species known in New Zealand and 18 of these occur mainly in subalpine regions. The different species are difficult to identify, as adult and juvenile forms may differ from one another and, where two species occur together, adult forms may give rise to swarms of hybrid plants. Only some of the more common and outstanding species are illustrated here.

Family: Epacridaceae

27. Curved-leaf grass tree, Mt Ruapehu, 1,400m (December)

Curved-leaf grass tree   *Dracophyllum recurvum*
Easily recognised by its often reddish-tinged, grey-green, recurved leaves, 1-4cm long, this grass tree attains a height of 60cm and is found on the central Volcanic Plateau, and Ruahine and Kaimanawa Mountains amongst alpine scrub and fellfields. Flowers, 6mm long, occur as stout spikes during January and February. This plant can spread over quite large areas.

Oliver's dracophyllum   *Dracophyllum oliveri*
A fine grass tree reaching a height of 2m and found throughout the South
Island in subalpine scrub near seepages, swamps and damp places. It also
grows in lowland places near swamps and along lake shores. Flowers 3-
4mm long occur from November to January.

28. A specimen of a typical grass tree, Oliver's dracophyllum, Mt Belle,
Homer (December)

29. Buds and flowers of Oliver's dracophyllum, Lake Manapouri
(November)

30. Specimen of the large grass tree, Canaan, Abel Tasman
National Park (November)

Large grass tree   *Dracophyllum traversii*
This, the largest of the grass trees, reaches a height of 10-13m. The habit
of growth of the younger trees is shown by the two on the right, while
the open habit of the older mature trees is shown by the fine specimen
on the left of figure 30. The large grass tree flowers during January and
February, the flower panicle being terminal on the branches. The leaves
are 30-60cm long, and the plant is found along the margins of subalpine
forest and in subalpine scrub from 800 to 1,400m. On the slopes of Mt
Arthur, behind Nelson, and in the Moor Park area of Abel Tasman National
Park there are almost pure stands of this grass tree. In the North Island
the very similar nei nei, *D. latifolium*, occurs in forests from sea level
to exposed subalpine ridges at 1,000m.

31. Totorowhiti in flower, Ruapehu (November)

32. Spreading grass tree showing habit and flower sprays, Mt Belle, Homer (December)

33. Mountain nei nei, showing spent flower panicle, near Boulder Lake (February)

Mountain nei nei   *Dracophyllum townsonii*
Very similar to the large grass tree but only attaining 6m in height.
Distinguished by its smaller leaves, 15-30cm long, and its flower panicles,
which are lateral on the branches as shown in figure 33. Found in subalpine
regions of southwest Nelson and north Westland.

Totorowhiti   *Dracophyllum strictum*
A small shrub attaining 70cm in height, typical in leaf form and flower
panicle of the broad-leaved forms of *Dracophyllum*. The leaves, 10cm
long by 12mm wide, taper to a hard, pungent apex. Flowers occur from
November to March, and the plant is found in damp scrub from sea level
to 1,000m altitude from Thames southwards.

Spreading grass tree   *Dracophyllum menziesii*
A much-branching, spreading shrub, often covering extensive areas and
seldom reaching 1m in height. Leaves, 7-20cm long and 1-2cm wide, are
often tinged with red. The flowers occur from December to February,
and the plant is found in damp alpine scrub or fellfields of the South
Island up to 1,400m altitude. Very extensive areas in the vicinity of the
Homer tunnel used to be covered by this plant.

34. Monoao with green spring foliage and flowers, Waipunga Gorge (December)

35. Monoao with brown winter foliage and spring buds just starting, Waipunga Gorge (October)

36. Needle-leaf grass tree with flowers, Mt Holdsworth (December)

Needle-leaf grass tree   *Dracophyllum filifolium*

A shrub or a small tree up to 2m high, rather similar to monoao but with leaves 6-16cm long, rather thread- or needle-like, with their apices three-sided. Found in subalpine and alpine scrub and fellfields in the southern half of the North Island only.

Monoao   *Dracophyllum subulatum*

A needle-leaved grass tree up to 2m high, with rigid, leathery, sharp-pointed leaves, 18-40mm long on young plants and 8-30mm long on mature plants. Flowers 2-3mm long occur from November to March. Monoao is found only in the North Island, from Thames south to the Ruahine Mountains, in scrub up to 1,100m altitude. During winter months the plant turns a deep bronze-brown, changing to fresh green as the spring growth appears. Monoao often occurs in pure stands over fairly large areas, and during the winter these areas of bronze-brown are quite striking. Large areas of many hectares used to exist between Taupo and the Kaimanawa Mountains but these, over the past decade, have been converted to farmlands and pine forests.

37. Turpentine scrub in flower, Mt Holdsworth (February)

38. Trailing grass tree in flower, near Lake Tekapo (December)

39. Bush rice grass, Tauhara Mountain, 800m (December)

Turpentine scrub *Dracophyllum uniflorum*
An erect shrub 1m high, with stout, pungent, needle-like leaves, 12-25mm long, having pubescent upper surfaces and ciliate margins. Flowers 5mm long occur singly at the apices of lateral branches during January and February. Found in alpine scrub, herbfields, fellfields and grasslands from the Kaimanawa Ranges southwards.

Trailing grass tree *Dracophyllum pronum*
A trailing or prostrate *Dracophyllum* with narrow, thick, leathery leaves, 5-12mm long and 1mm wide, swollen at their apices and with minutely serrated margins. Flowers, about 4mm deep and 4-5mm wide, occur singly along the branchlets. Found from subalpine scrub to fellfields amongst the South Island mountains, mainly east of the main divide.

Bush rice grass *Microlaena avenacea*
Common in lowland forest, this grass is also often found in subalpine forest and scrub in shady, damp places. Readily recognised by its soft, broad leaves with conspicuous midribs, its clumps reach a height of 1.3m. The slender flower panicles, up to 60cm long, occur during December and January.

Family: Gramineae

THE GENUS COPROSMA

There are 45 species of *Coprosma* in New Zealand and many of these
are found in subalpine and alpine regions; some, in fact, occur only in
high mountain country. Coprosmas are very variable in form and habits;
the male and female flowers are quite distinct and are borne on separate
plants. Coprosmas belong to the family Rubiaceae and they all produce
large, usually highly coloured drupes in the autumn. Most species show
domatia (small pits) on the lower surfaces of their leaves. These are usually
lying in the angles between the midrib and a lateral vein. A number of
species occur in subalpine and alpine scrub, and a selection of these is
shown on the following pages.

Family: Rubiaceae

40. Sprawling coprosma with ripe drupes, Volcanic
Plateau (February)

Sprawling coprosma   *Coprosma cheesmanii*
A sprawling or prostrate shrub with divaricating, pubescent branches often
forming very tight, compact bushes. The narrow leaves are 8-11mm long
by 1-2mm wide, and the drupes, 6-7mm across, which ripen during February
and March, are produced in great profusion. Found in subalpine scrub
tussock and grasslands, bogs and fellfields from East Cape southwards.

Variable coprosma   *Coprosma pseudocuneata*
A stoutly branched species up to 3m high, of variable form and habit,
and found from East Cape southwards in subalpine forest or scrub and
alpine bogs, grasslands or herbfields. The thick, leathery leaves are 5-
20mm long and 2-6mm wide. The drupes, 5-6mm long are produced in
profusion, and easily dislodged when mature; they ripen from February
to May and persist on the plants in alpine situations until December or
even January.

41. Variable coprosma with ripe drupes, Cupola River (April)

42. Variable coprosma close up to show leaf arrangement and drupes,
    Arthur's Pass (April)

43. The reticulate coprosma with mature drupes, Arthur's Pass (April)

The reticulate coprosma   *Coprosma serrulata*

Readily identified by its large, markedly reticulated leaves 4-7cm long, and its white bark that falls in flakes, this coprosma is common in subalpine and alpine scrub, tussock grasslands and herbfields throughout the South Island. Shrubs grow to 1m high with stout, spreading branches. The drupes are 7-8mm long and make a magnificent display in the autumn.

Needle-leaved mountain coprosma   *Coprosma rugosa*

An erect, somewhat rigid shrub up to 3m high, often with divaricating branches clothed with reddish bark. The narrow leaves are thick and pointed, 10-14mm long and 1-1.5mm across. Note in figure 45 the typical coprosma female flowers in which the styles are the conspicuous parts and look like twigs at first glance. The translucent drupes, 6-8mm long, ripen during April and May and can persist on the plants throughout the winter.

44. Needle-leaved mountain coprosma with ripe drupes, Arthur's Pass (April)

45. Needle-leaved mountain coprosma with female flowers and drupes, Volcanic Plateau (October)

Mingimingi   *Coprosma propinqua*
A common coprosma of wet lowland forests, Mingimingi can also be met
with in subalpine forest and lower subalpine scrub, especially along
streamsides and damp, shady banks. I have come across it in the upper
Hollyford Valley and along streambanks of the Volcanic Plateau. A shrub
or small tree, it can reach a height of 6m. The leaves, 10-14mm long
by 2-3mm wide, have prominent veins and reticulation. The male flowers,
3mm across (fig. 46), are typical of these flowers in coprosmas, and the
drupes, 8mm long, vary from yellow to pale or deep blue.

46. Male flowers of mingimingi, Upper Waipunga Valley (October)

47. Mingimingi with deep blue drupes, Upper Waipunga Valley (March)

48. Pale blue drupes of mingimingi, Molesworth (April)

Stinkwood   *Coprosma foetidissima*
A rather open-branched shrub, 3-4m high, found in lowland and subalpine forest and scrub all over New Zealand. Stinkwood is readily recognised by its vile odour, like the smell of rotten eggs, which is emitted when any portion of the plant is crushed, broken, or even brushed against. The leaves are 15-30mm long and 14-20mm wide, and the drupes, 7-10mm long, ripen during April and May.

49. Stinkwood with drupe, Mt Holdsworth (June)

Springy coprosma   *Coprosma colensoi*
A slender, springy coprosma, reaching 2m in height and with small, narrow, obovate to elliptic leaves, 15-20mm long by 5-8mm wide, on hairy petioles 2-3mm long and with the midrib strongly evident. The striking, oblong-shaped drupes, 6mm long, occur singly or in pairs and ripen during May and June. Springy coprosma is found in subalpine forest and scrub from the Coromandel Peninsula southwards. *C. colensoi* includes the species previously known as *C. banksii*.

50. Springy coprosma with drupes, Mt Holdsworth (June)

51. Wavy-leaved coprosma with drupes, Mt Egmont (May)

52. Foliage of small-leaved pepper tree, Boulder Lake (February)

Small-leaved pepper tree   *Pseudowintera traversii*
A closely branched, erect shrub up to 2m high, with a rough, often wrinkled
or warty, acrid bark and close-set, thick, overlapping leaves 10-25mm long.
Flowers and fruits are similar to those of horopito, and the plant is found
in subalpine forest and scrub of northwest Nelson.

Family: Winteraceae

Wavy-leaved coprosma   *Coprosma tenuifolia*
A slender, sparingly branched shrub up to 5m high with broad, ovate
leaves, rather membraneous, 4-10cm long with wavy margins and
conspicuous veins. The midrib is hairy above, and the leaves are on thin,
hairy petioles 10-25mm long. The ovoid drupes, 7-8mm long, ripen during
April and May, and the plant is found from Te Aroha Mountain southwards
to the Ruahines. It is common in the subalpine scrub of Mt Egmont to
about 1,250m altitude and in the vicinity of Lake Taupo.

Family: Rubiaceae

53. Coloured horopito leaves, Moor Park Track, Abel Tasman National Park (November)

54. Horopito leaves, upper surfaces showing normal blotching, Mt Holdsworth (December)

Horopito   *Pseudowintera colorata*

An erect, bushing shrub or a small tree up to 10m high, noted for its blotch-marked and often beautifully coloured leaves, about 6cm long and 3cm wide, with bluish undersides and strong, pungent taste; the leaves are on petioles up to 1cm long. Found throughout New Zealand in subalpine scrub and lowland to subalpine forests. The plant often forms thickets in open forests and beautifully rounded bushes in open scrub. The flowers, about 1cm across, are highly aromatic and occur from September to November. The fruit is a black berry, which ripens during February and March.

Family: Winteraceae

55. Horopito leaves showing bluish-coloured undersides and a branchlet with flowers, Mt Holdsworth (December)

Pepper tree   *Pseudowintera axillari.*

A shrub or small tree up to 8m high, occurring commonly in lowland and subalpine forests throughout New Zealand. Pepper trees often form thickets where the forest begins to open out towards the subalpine bush line, and plants often extend into the region of subalpine scrub. The shining aromatic leaves, 6-10cm long, have dull green or only faintly bluish undersides. They are distinctly pungent to taste. The axillary flowers, similar to horopito flowers, are 10-12mm across and occur from September to December. The red berries are ripe by the following May or June, and flower buds for the next year are usually well formed by this time. (See also fig. 59.)

Family: Winteraceae

56. Pepper tree berries, Mt Holdsworth (June)

57. Mountain wineberry with flowers, showing habit of growth, Outerere stream, base of Mt Tongariro (December)

Mountain wineberry   *Aristotelia fruticosa*

A much-branched, somewhat tangled but variable shrub, reaching 2m high, with ovate or ovate-oblong leaves 5-15mm long. The flowers, which occur from October to December, are in small clusters, and the berries, 3-4mm in diameter, ripen from February till April. Found in subalpine scrub, grasslands and fellfields throughout New Zealand.

Family: Elaeocarpaceae

58. Mountain wineberry, with fruits, Molesworth (March)

59. Pepper tree leaves showing green undersides (*see* fig. 56) Mt Holdsworth (October)

60. Mountain five-finger with male flower buds opening, Mt Egmont (May)

61. Shrubby kohuhu showing growth habit and flowers, Wharite Peak (December)

Shrubby kohuhu   *Pittosporum rigidum*

A stiff, densely branched shrub, up to 3m high, with stout, non-interlacing branchlets bearing small, thick, leathery leaves 8-10mm long and 5-8mm across. Found in subalpine scrub up to 1,250m, from the Kaimanawa Mountains to Southwest Nelson. Small axillary flowers occur from September to February.

Family: Pittosporaceae

Mountain five-finger   *Pseudopanax colensoi*

A shrub reaching 5m high and found in subalpine forest and scrub throughout New Zealand. Male and female flowers occur on separate plants, from May to October, and are strongly sweet-scented. The round, flat, black seeds occur in large bunches, which ripen about March of the following year. The leaflets are 5-15cm long, thick, smooth and glossy with serrated margins.

Family: Araliaceae

53. Flowers of Dall's pittosporum, near Boulder Lake (December)

Dall's pittosporum   *Pittosporum dallii*

A spreading shrub or small canopy tree, up to 6m high, found only in the vicinity of Boulder Lake and a few other isolated areas of northwest Nelson. The coarsely serrated leaves up to 10cm long are crowded towards the tips of the branchlets, and the flowers, which arise as terminal umbels from November to January, have a strong, delightfully sweet scent.

Family: Pittosporaceae

Black mapou   *Pittosporum tenuifolium* subsp. *colensoi*

A tree up to 10m high, found in subalpine forest and scrub from the Raukumara Ranges southwards to Stewart Island. The dark-green leaves are pale below and vary from 5 to 10cm long and 2 to 4cm wide. Axillary flowers occur during November and December.

Family: Pittosporaceae

62. Spray of black mapou with flowers, Kaimanawa Mountains (November)

HEBES

The genus *Hebe* includes about 80 species of evergreen shrubs found in New Zealand. Many of these occur only in mountain regions, where they are found in scrub, alpine bogs, tussock lands and rocky places. They all belong to the family Scrophulariaceae.

64. Flowers of Ruapehu hebe, Volcanic Plateau (February)

65. Flowers of Tararua hebe, Mt Holdsworth (February)

66. Spray of northwest Nelson hebe showing branchlets and flowers, Mt Robert, Nelson Lakes (January)

Northwest Nelson hebe *Hebe coarctata*

A spreading shrub up to 1m high but usually somewhat less; the branches are arching and tending to be decumbent, with numerous branchlets arising along the upper sides only. Found in Northwest Nelson south to Mt Robert, at Lake Rotoiti and the Brunner Range; often on the edge of the beech forest.

Ruapehu hebe *Hebe venustula*

An erect hebe up to 1.5m high, occurring in subalpine scrub from Mt Hikurangi south to Mt Ruapehu. Flowers occur in profusion as racemes 3-4cm long from December to February.

Tararua hebe *Hebe evenosa*

Found on the Tararua Mountains, this hebe with stout, widely spreading branches reaches 2m in height. The smooth, thick leaves are 15-20mm long. The flowers, which occur during January and February, are crowded on lateral spikes around the tips of the branchlets.

67. Close view of the stem of the ochreous whipcord showing appressed leaves

68. The ochreous whipcord with flowers, Mt Peel, Nelson (December)

Ochreous whipcord   *Hebe ochracea*

A rigid, spreading whipcord hebe, up to 30cm high with glossy, olive-green branchlets becoming ochreous coloured towards their tips. The thick leaves, 1-1.5mm long, have fine, hairy margins and are keeled and closely appressed to the stems (fig. 67). Flowers occur as terminal spikes of up to 10 flowers each during December and January.

Mountain koromiko   *Hebe subalpina*

A densely branched shrub up to 2m high with thick, shining, leathery leaves, 2.5cm long and 5-8mm wide, which produce conspicuous leaf scars as they fall. The flower stalks are longer than the leaves and arise laterally during December and January. Found from Nelson to Fiordland in rocky places amongst subalpine and alpine scrub, more especially in areas of higher rainfall. (See also page 192.)

69. Flowers of Mountain koromiko, Arthur's Pass (December)

70. Branch of varnished koromiko with flowers, Mt Peel, Nelson
(December)

Varnished koromiko   *Hebe vernicosa*
A semi-prostrate or erect shrub found in subalpine scrub and fellfields
throughout Marlborough and Northwest Nelson. The small leaves, 15mm
long, have a varnished appearance and their petioles twist so that although
they do arise opposite and alternate on the branches, they do not appear
to do so. Flower heads up to 5cm long arise laterally towards the tips
of the branches. This hebe also grows in the beech forests in the Nelson
Lakes region from sea level to the upper limits of the forests, where it
reaches a height of 1m.

Dwarf whipcord   *Hebe hectori* var. *demissa*
Hebe hectori is an erect, rigid, branching shrub, 10-75cm high, but the
variety *demissa* is a low-spreading shrub, 10-15cm high, with circular
branchlets 2mm in diameter clothed with convex appressed leaves, each
with a small, sharp-keeled tip. The parent species, *H. hectori,* is generally
larger than the variety *demissa,* with the branchlets 2-3.5mm in diameter,
the leaves 2-2.5mm long and the thick branchlets marked by the scars
of fallen leaves. Found in wet subalpine scrub and grasslands among the
mountains of South Canterbury and Otago.

Club-moss whipcord   *Hebe lycopodioides*
An erect, rigid, branching shrub, up to 1m high, very similar to the pumice
whipcord (fig. 74) and found in subalpine scrub and grasslands among
the mountains of Marlborough and Nelson and southwards along the
Southern Alps to the vicinity of Lake Wakatipu. Branchlets 2-3mm in
diameter are four-angled and clothed with thick, appressed, concavo-convex
leaves, 1.5-2mm long, each with an apical blunt cuspor spine (fig. 72).
Flowers 5-8mm across occur as spikes during December and January.

71. Close view of stem of dwarf whipcord

72. Close view of four-angled stem of club-moss whipcord, with appressed leaves

73. Club-moss whipcord with flowers and buds, Garvie Mountains (December)

Pumice whipcord   *Hebe tetragona*
A stiff, erect, branching shrub, reaching 1m high, with rigid four-sided
branches bearing very thick, broad, scale-like, leathery leaves 2-3mm long.
The small, white flowers, 6-8mm across, are terminal on the branches
and appear from December to February. Found in subalpine scrub,
grasslands and fellfields of the North Island, from Mt Hikurangi south
to the Volcanic Plateau and the Ruahine Mountains. This photo shows
clearly the four-sided structure of the stem, and the thick leaves.

74. Close view of stem of pumice whipcord, Mt
Ruapehu (December)

Cypress-like hebe   *Hebe cupressoides*
A finely and densely branched, rounded shrub, up to 2m high, found
on river flats and terraces east of the divide, from Marlborough to Otago,
up to about 1,500m altitude. The branchlets are about 1mm through and
the young leaves are appressed, 1-1.5mm long, but spreading as they age.

Creeping matipo   *Myrsine nummularia*
A prostrate, rambling shrub with trailing stems up to 50cm long and reddish-
brown bark. The thick, round-to-oval leaves about 1cm long are dotted
with glands, and vary from dark green through golden-brown to reddish-
brown or purple in colour. The plant occurs in rocky alpine scrub and
fellfields and is very common along the edges of screes up to 1,600m,
where its creeping habit helps to stabilise these rapidly eroding areas. Minute
flowers occur from October to February, and these beautiful berries ripen
during April and May.   Family: Myrsinaceae

75. Branchlets and flowers of *Hebe cupressoides*,
Marlborough, (December)

76. Creeping matipo with berries, Arthur's Pass (April)

77. Weeping matipo plant showing habit, Lewis Pass (January)

Weeping matipo   *Myrsine divaricata*
A shrub up to 3m high with rough, dark-brown bark, many interlacing
branches and drooping branchlets that are pubescent when young. The
broad, obovate leaves are 5-15mm long and dotted with glands. The minute
flowers occur singly or in clusters of twos and threes along the branchlets
from June to November, and the beautiful deep-blue or mauve-coloured
berries, 4-5mm across, ripen between August and the following May.
Weeping matipo is found in damp subalpine and alpine scrub and fellfields
as well as in damp lowland forest throughout New Zealand.
                                                    Family: Myrtaceae

78. Flowers of weeping matipo, Upper Travers Valley (August)

79. Fruits of weeping matipo, Upper Travers Valley (April)

80. Haumakaroa with juvenile leaves, Moor Park Track, 1,000m, Abel Tasman National Park (November)

81. Haumakaroa with berries, Upper Otira Gorge (April)

82. Mountain cabbage tree in flower, Mt Egmont
(December)

Mountain cabbage tree   *Cordyline indivisa*
A tall, single-stemmed agave up to 4m high, found in subalpine forest
and scrub of the North Island and Westland between 500 and 1,100m.
The long, very broad leaves and the flower head that hangs beneath the
leaves distinguish *C. indivisa* from other cabbage trees. The flower occurs
during November and December. The related *C. banksii* usually has several
stems, narrower leaves and the more open-type flower head above the
leaves. *C. banksii* occurs up to 1,000m.

Family: Agavaceae

Haumakaroa   *Pseudopanax simplex*
A shrub or small tree, up to 8m high, which passes through two juvenile
stages characterised by distinct leaf forms, the second of which, with leaflets
5-8cm long and petioles 6cm long, is shown in figure 80. First-stage juvenile
leaves have longer petioles (up to 10cm) long with longer leaflets up to
15cm long. Adult plants have mostly single leaves but trifoliate leaves
can occur. Small greenish flowers arise as umbels with female flowers
above and males below from November to January. The rather flattened
seeds ripen from April onwards through the winter, becoming dark brown
to black when fully mature.

Family: Araliaceae

83. Female cones of mountain pine, Lewis Pass (March)

Mountain pine   *Halocarpus bidwilli*
A spreading-to-erect, dome-shaped shrub, up to 3.5m high and 5-6m across
with the lower branches rooted. The young leaves are thick and spreading,
up to 10mm long, but adult leaves are short and appressed to the stem.
Mountain pine is found in subalpine scrub and exposed rocky and boggy
places throughout the mountains of New Zealand.

Family: Podocarpaceae

84. A typical mountain pine, 900m, Lewis Pass (January)

85. Pigmy pine with male strobili, Arthur's Pass (November)

86. Pigmy pine with seeds, Arthur's Pass (April)

Pigmy pine   *Lepidothamnus laxifolius*

The smallest pine in the world, often fruiting in the juvenile form, this prostrate creeping plant is common in subalpine scrub and fellfields from Tongariro National Park southwards. Juvenile leaves are spreading and up to 12mm long; adult leaves only 1-2mm long are appressed to the stems. The plant forms dense turfs and mats trailing over stones when mature. Male strobili appear during November and December, and fruits ripen during March and April.

Family: Podocarpaceae

87. Plant of mountain flax with old seed stalks and new season's flower buds, Mt Ruapehu (December)

88. Normal orange-red flowers of mountain flax, Volcanic Plateau (November)

89. Red variety of mountain flax flowers, Volcanic Plateau (November)

Mountain flax   *Phormium cookianum*

Sometimes also called 'coastal flax' because it grows equally well in exposed coastal situations, this agave is one of the commonest plants in New Zealand. It is found in the mountains in rocky places amongst subalpine and alpine scrub and fellfields up to altitudes of 1,600m, flourishing in either dry or wet places. Leaves are 1-2m long and 3-7cm wide. Flowers on stalks up to 2m high occur from October to January and are normally orange-red, but deeper red and lighter, yellowish-green flowers also occur. The seed pods always hang downwards from the flower stalk.

Family: Agavaceae

SNOWBERRIES

Snowberries belonging to the genera *Gaultheria* and *Pernettya* are four throughout the New Zealand mountains from subalpine scrub to the high herb and fellfields. All species have edible fruits, and records exist of peop who have been sustained by eating them. They form attractive, erect spreading shrubs that can be prolifically covered with fruits in season.

90. Snowberry flowers, Tauhara Mountain, Taupo (November)

91. White snowberries, Mt Holdsworth (February)

92. Red snowberries, Mt Holdsworth (February)

Snowberry   *Gaultheria antipoda*

An erect or spreading shrub, 30cm to 2m high, with the branchlets densely or sparsely clothed with black or yellowish bristles mingled with a fine, silvery down. The thick, leathery leaves, 5-15mm long and up to 15mm wide, have conspicuous veins and bluntly serrated margins.

Snowberry is a very variable plant found throughout New Zealand in subalpine and alpine scrub, fellfields and rocky places. The small flowers, which may be red as well as white, occur from November to February. The fruit is a capsule, 12mm across, usually embedded in the fleshy calyx lobes and may be white, pink or deep red. Snowberry hybridises freely with prostrate snowberry, producing plants with leaf forms and fruits intermediate between the two species.

Family: Ericaceae

93. Tall snowberry plant in flower, Mt Holdsworth (December)

Tall snowberry   *Gaultheria rupestris*
An erect, branching shrub, reaching 1m high, found in subalpine scrub,
grasslands and rocky places in the Ruahine and Tararua Ranges and in
the mountains of Nelson. Flowers occur as terminal or subterminal racemes
up to 4cm long from November to March.

Family: Ericaceae

94. Tall snowberry flowers, Mt Holdsworth (December)

TUSSOCK GRASSLANDS
AND CARPET GRASSLANDS

Mountain tussocks and carpet grasses of the family Gramineae clothe vast areas of the rolling mountain slopes above the bushline throughout New Zealand, forming meadows that extend to an altitude of 2,000m. In amongst the tussocks in a healthy meadow the ground is carpeted with grasses, sedges, creeping mat plants, herbs, bluebells, bidibidi, wild spaniards and orchids. Some shrubby plants often occur amongst tussocks, and tussock meadows often merge with herbfields.

95. Narrow-leaved snow tussock field, *Chionochloa pallens*, Cupola Basin (April)

Narrow-leaved snow tussock   *Chionochloa pallens*
Densely tufted, brownish-green tussocks, 60cm to 2m high, found abundantly from Mt Hikurangi and Mt Egmont southwards, up to elevations of 1,700m. This is probably the commonest tussock of the South Island mountains.

Family: Gramineae

96. Plants of narrow-leaved snow tussock, *Chionochloa pallens*, with seeding glumes, Cupola Basin (April)

97. Narrow-leaved snow tussock in seed, Mt Holdsworth (February)

98. Hunangamoho in flower, Hollyford Valley (January)

Hunangamoho *Chionochloa conspicua*
A large tussock, up to 2m high, found in alpine grasslands, subalpine and lowland grasslands, and natural forest clearings throughout New Zealand. The strongly nerved, flat leaves, 45cm to 1.2m long and 6-8mm wide, are usually softly hairy along their margins. Handsome flower panicles up to 45cm long occur from November to January.

Family: Gramineae

99. Red tussock, Lewis Pass (January)

Red tussock   *Chionochloa rubra*
Large, dense, brownish-green tussocks up to 1.6m high, found from East
Cape and Mt Egmont southwards. Although found in alpine grasslands
to altitudes of 1,900m, red tussock also occurs down to sea level. It is
abundant throughout the mountain regions of the South Island.

Family: Gramineae

Carpet grass   *Chionochloa australis*
A low-growing grass forming tight mats, often covering extensive areas
of steep alpine slopes in Nelson and Canterbury up to heights of 2,000m.
After snow, carpet grass often lies down the slope, when it may be extremely
slippery and dangerous to traverse. Always a lush deep green, it adds
freshness to the mountain regions. Flowers are produced in December
and January, followed by silvery seed heads, which glisten attractively
in the sunshine (fig. 101).

Family: Gramineae

00. Patch of carpet grass in a tussock field, Cupola Basin (April)

01. Carpet grass in seed, Douglas Ridge, Boulder Lake (January). *Celmisia incana* in the foreground

102. Silver tussock in flower, Tauhara Mountain (January)

Silver tussock   *Rytidosperma viri*
A tufted grass with fine, silvery-green leaves forming compact tussock
up to 1m high. Flowers on stems up to 60cm high occur during Decembe
and January. Silver tussock occurs throughout New Zealand up to 1,400ₘ
in subalpine meadows and exposed, stony places.

Family: Gramineae

Tall mountain sedge   *Gahnia rigid*
Leaves more or less erect, with long, thin, drooping tips and inrolled
cutting margins. Erect, rigid flower stems, 60cm-2m high, occur durin
January and February. Found in subalpine regions of the mountains c
Nelson and Westland.

Family: Cyperacea

Tall tufted sedge   *Gahnia procer*
A tufted sedge up to 1m high with slender, curving flower stems 50cm
to 1m long, producing large panicles of flowers during October, Novembe
and December. The seed, which is an ovoid, smooth, shining, brown nut
6mm long, ripens during February and March. Found from North Cap
southwards to Westland, it occurs in subalpine grasslands from abou
1,000m down to sea level.

Family: Cyperacea

103. Tall mountain sedge in flower,
track to Boulder Lake (January)

04. Tall tufted sedge in flower,
slopes of Mt Ruapehu
(December)

105. Scarlet bidibidi in fruit, Volcanic Plateau (January)

106. Blue mountain bidibidi in flower, Boulder Lake (February)

Scarlet bidibidi   *Acaena microphylla*
A prostrate, branching herb that when in flower forms bright-red patches
up to 75cm across. The hairless leaves are up to 3cm long, and the flower
heads, 2.5cm in diameter, occur from December to February.
Found in grasslands, riverbeds and herbfields of the Volcanic Plateau
around Tongariro National Park, up to 1,100m altitude.
Family: Rosaceae

Blue mountain bidibidi   *Acaena inermis*
A spreading, prostrate herb with soft, bluish-coloured, five- to seven-foliate
leaves up to 5cm long. Leaves are bluish-green in the shade but turn bronze
in hot sun. This plant occurs in grasslands and fellfields and along the
edges of screes in the South Island to elevations of 1,600m. Flower heads
are unarmed and occur from December to February.
Family: Rosaceae

107. Red bidibidi in fruit, Volcanic Plateau (November)

Red bidibidi   *Acaena novae-zelandiae*
A silky-haired herb with creeping and rooting stems up to 1m long, often
forming large patches in subalpine grasslands. The globose flower heads
occur from September to December, and the seeds, which ripen from
February on, attach themselves readily to clothing and the fur of animals.
Found throughout New Zealand in lowland and subalpine grasslands and
open, dry places.
Family: Rosaceae

108. Plants of the odd-leaved orchid, Lewis Pass
(December)

109. Flowers of the odd-leaved orchid, Lewis Pass (December)

110. The grassland orchid, Volcanic Plateau (February)

111. The grassland orchid, Ruahine Mountains (April)

Grassland orchid *Orthoceras strictum*

A slender, rigid orchid, 20-60cm high, found in both subalpine and lowland, dry, open grasslands and herbfields from Northland south to Nelson. Flowers occur from December to April as spikes 2-22cm long; each spike consists of 3-12 flowers, which vary considerably in their colours between plants. Two variations are shown here.

Family: Orchidaceae

Odd-leaved orchid *Aporostylis bifolia*

A two-leaved orchid with one leaf much longer and broader than the other, the small leaf being 1-2cm long, the larger 4-7.5cm long. The flower, 12-14mm long, may be either white or pink, and is borne on a hairy stem up to 10cm high during December and January. This orchid is found throughout New Zealand up to 1,400m altitude in damp grasslands, damp places in herbfields and along the edges of bogs.

Family: Orchidaceae

112. Lake Tekapo epilobium in flower, Lake Tekapo (December)

Lake Tekapo epilobium   *Epilobium rostratum*
A spreading herb with stiff, erect stems 4-15cm high and leaves 5-8mm
long, which is found in subalpine grasslands and alongside streambeds
from Arthur's Pass to Lake Wakatipu. It is common round Lake Tekapo.
Flowers 3-5mm across occur from December to February.

Family: Onagraceae

Onion-leaved orchid   *Microtis unifolia*
A variable orchid with stout or slender, fleshy stems, 7-60cm high, and
a single, rounded, onion-like leaf that is usually longer than the flower
spike. Flowers occur from October to February on spikes 1-15cm long,
with few or many flowers, closely or sparsely spaced. Found throughout
New Zealand up to 900m, in tussock grassland and sometimes in herbfields.

Family: Orchidaceae

Bedstraw   *Galium perpusillum*
A prostrate, spreading and rooting, perennial herb forming close-set patches
up to 30cm across, found commonly in damp situations between tussocks,
in fellfields near streams, and in partially shaded rocky places. Three species
of *Galium* are indigenous to New Zealand. All are found in lowland regions
but *G. perpusillum* reaches to higher altitudes and I have found it at
1,400m in Cupola Basin. The branchlets are ascending, bearing minute
leaves 1.5-3mm long, arranged in whorls of four around the stems. The
small flowers, 2mm across, occur in profusion from December to March
and are either axillary or terminal.

Family: Rubiaceae

113. Flower spike of onion-leaved orchid, Volcanic
Plateau (December)

114. Bedstraw in flower, Cupola Basin, 1,550m (December)

Rimu-roa *Wahlenbergia gracilis*
A herb, usually perennial, up to 40cm high with woody stems at the base.
The slender, upright stems, much branched, bear variable leaves 1-4cm
long, the upper ones usually linear or lanceolate. Flowers, up to 2cm across
on long stems, occur from September to April. Rimu-roa is found in both
subalpine and lowland grasslands throughout New Zealand.

Family: Campanulaceae

115. Rimu-roa flower, Volcanic Plateau (December)

Bog daisy *Celmisia graminifolia*
A tufted daisy found commonly in damp places in grasslands, scrub,
herbfields and fellfields throughout the New Zealand mountains. The
pointed leaves, 5-20cm long by 4-5mm wide, are smooth above but clothed
below with dense felt, and the margins are usually rolled slightly inwards.
Flowers 1-1.5cm across occur during December and January.

Family: Compositae

Grassland daisy *Brachycome sinclairii* var. *pinnata*
Grassland daisy is a creeping herb forming patches up to 30cm across,
with leaves up to 8cm long rising from the rootstock; the leaves, variably
pinnatifid, are distinctly so in variety *pinnata*. Flower stalks, up to 30cm
high, each bearing a single flower 1.5-2.5cm across, occur from November
to January. This daisy is found in alpine grasslands and herbfields from
East Cape southwards up to altitudes of 2,100m.

Family: Compositae

116. Bog daisy, Arthur's Pass (January)

117. Grassland daisy in flower, Old Man Range (December)

118. Prostrate snowberry flowers, Mt Ruapehu (November)

119. Prostrate snowberry hybrid with red berries, Outerere Gorge, Mt Tongariro (May)

120. Prostrate snowberry hybrid with white berries, Volcanic Plateau (February)

Prostrate snowberry   *Pernettya macrostigma*
A much-branched, prostrate or straggling shrub up to 20cm high, forming patches in subalpine and alpine grasslands, fellfields and rocky places throughout New Zealand. It hybridises with *Gaultheria* species and figures 119 and 120 are of a hybrid with *Gaultheria depressa* var. *novae-zelandiae*. The small, solitary, axillary flowers, about 3mm long, appear from November till January. The large, fleshy berries, 4-7mm across, are each provided with a persistent, fleshy calyx matching the berry in colour. Berries vary from white to deep red and ripen from December to May. The leaves are thick and leathery, 5-10mm long and 1-4mm wide, with *Gaultheria* having delicately serrated margins.

Family: Ericaceae

Little mountain heath   *Pentachondra pumila*
A rather slender, branching, dwarf shrub, forming patches up to 40cm
across and found throughout New Zealand in subalpine and alpine
grasslands, fellfields, along the edges of alpine bogs and in alpine rocky
places up to 1,300m. It also occurs sparingly in lowland grasslands and
sand-dune country. The crowded leaves, up to 5mm long, vary from oblong
to broad elliptic, and the very small, solitary, axillary flowers occur near
the tips of the branches from November to February. The large berry,
6-12mm across, which ripens from December till April, frequently has
the style or the corolla of the flower remaining attached to it, as shown
in this photograph. Berries and flowers normally occur together.

Family: Epacridaceae

121. Little mountain heath with flowers and berries, Jack's Pass (January)

Creeping pohuehue   *Muehlenbeckia axillaris*
A prostrate, spreading shrub with woody stems, forming patches up to
1m across in subalpine grasslands, riverbeds, and stony or rocky places
throughout New Zealand. Leaves 3-5mm long on slender petioles 2-3mm
long. The flowers, 4mm across, which are either solitary or in pairs, occur
from November to April, filling the evening air with their rich, sweet scent.
The black seeds, each about 3mm long sitting in a white cup, ripen from
December on.

Family: Polygonaceae

122a. Creeping pohuehue spray with flower (close-up), Haast Pass (January)

22b Close-up of pohuehue fruit, Haast Pass (January)

WILD SPANIARDS

Wild spaniard or speargrass is the popular name given to plants of the genus *Aciphylla,* which occur commonly throughout New Zealand from sea level to 1,850m. Thirty-nine species of these plants are known, and are characterised by their pinnate, sword-like or spear-like, pointed, rigid leaves, capable of inflicting a painful wound on the unwary. They are found in grasslands, herbfields, alpine and subalpine scrub and rocky places.

Family: Umbelliferae

123. The horrid spaniard in flower, Mt Belle, Homer (December)

Horrid spaniard   *Aciphylla horrida*

A large, stout *Aciphylla* forming clumps up to 1m high with stiff, sword-like, once-pinnate leaves up to 80cm long, having rigid, pungent tips. Flowers borne along stalks 1.5m tall occur during December and January and are most spectacular. The plant is found in alpine scrub on the western side of the main mountains in the South Island from Arthur's Pass to Fiordland. A similar but slightly smaller plant, *A. ferox,* is found in the mountains of Nelson and Marlborough.

124. Common speargrass, Lake Pounui (November)

Common speargrass *Aciphylla squarrosa*

A large *Aciphylla* forming dense tussocks 1m high, and bearing three-pinnate leaves 1m long; the leaves are somewhat bluish, rigid, with serrulate or crenulate margins and midrib, and with pungent tips. The flower stalk, 60cm to 2m high, bears faintly sweet-scented flowers from November to January. Common speargrass occurs in subalpine grasslands from East Cape to the Wellington region and often descends to lower levels.

125. The giant spaniard in flower, Tasman Glacier Moraine (December)

The giant spaniard   *Aciphylla scott-thomsonii*
An *Aciphylla* forming huge tussocks that bear flower heads 1m long on stalks up to 4m tall during December and January. The bluish-green, one- to two-pinnate leaves, up to 1.5m long, have faintly serrate margins and hard, yellowish, pungent tips. Found in subalpine scrub and fellfields from the Mt Cook region southwards.

Wild spaniard   *Aciphylla colensoi*
A large *Aciphylla* up to 1m high with generally distinctly bluish-coloured, rigid, sword-like leaves. These are usually once-pinnate, but sometimes at the bases two-pinnate, 30-50cm long, with pungent tips. The flower stalk rises 2.5m like a pole from the centre of the plant, and the sweet-scented flowers are borne along the apical third of the stalk during November and December. Seeds ripen during February and March, and the plant occurs from East Cape to Canterbury in alpine and subalpine grasslands and herbfields.

126. Wild spaniard with mature seed heads, Mt
     Holdsworth (February)

127. Portion of a typical seed head of a wild spaniard, *A. colensoi*,
     showing seeds and protecting pungent 'needles', Mt Holdsworth
     (February)

128. Golden spaniards growing with tussocks in
a rock cleft, Cupola Basin, 1,550m (April)

Golden spaniard   *Aciphylla aurea*
A similar plant to *A. horrida*, reaching a height of 1m with rather fan-
shaped, stiff, pungent leaves up to 70cm long. The three to four leaflets
per leaf have faintly crenulate or serrate margins. Flower stems are stout
and grooved, bearing large, golden-coloured flower heads up to 75cm long,
which appear during December and January. Found in tussock lands and
herbfields amongst the mountains of eastern Nelson. The plant shown
in figure 128 bears two mature seed heads.

Tararua speargrass   *Aciphylla dissecta*
A small *Aciphylla* up to 40cm high, with leathery, flexible, pointed two-
to three-pinnate leaves 30cm long, and flower stalks 30cm high occurring
during November and December. Found only in grasslands of the Tararua
Ranges and easily mistaken for an *Anisotome* plant.

Feathery spaniard   *Aciphylla squarrosa* var. *flaccida*
This form of *A. squarrosa* is somewhat soft and delicate, with flower
heads up to 1.5m high and with two- or three-pinnate leaves about 80cm
long. Flower heads appear during December and January. Restricted in
distribution to damp places in subalpine and alpine scrub in the mountains
of the North Island.

129. Tararua speargrass in flower, Mt Holdsworth
(December)

130. The feathery spaniard in seed, Mt Holdsworth (February)

Armstrong's speargrass   *Aciphylla montana*
This small speargrass is found in the mountains of the South Island from
the Harris Mountains northwards to the Arrowsmith Range. The leaves
are pungent, thick and leathery, up to 30cm long on petioles up to 9cm
long; the flower stalks are up to 60cm high, with the male flower more
open than the female. A similar-looking speargrass, *A. lyallii,* is found
in the mountains of Fiordland.

131. Armstrong's speargrass in flower, a male plant growing near Godley
River, Lake Tekapo (December)

Pigmy speargrass   *Aciphylla monroi*
A small, tufted *Aciphylla* with four to six pairs of leaflets per leaf. The
leaves are soft and somewhat flexible but provided with pungent tips
and are 8-10cm long and 3-4cm wide. The flower stalks reach 30cm high
and flowers occur during December and January. Found in alpine and
subalpine grasslands, herbfields and rocky places up to 1,700m among
the mountains of Nelson, Marlborough and North Canterbury.

False spaniard   *Celmisia lyallii*
A rigid, tufted *Celmisia* that, with its long, narrow, stiff, pungent-tipped
leaves, is easily mistaken for an *Aciphylla* when not in flower. The leaves
are 20-45cm long by 6-9mm wide, with smooth upper surfaces and grooved
lower surfaces, more or less clothed with white, satiny hairs. Flowers 2.5-
5cm across are borne on slender, woolly stalks, 15-30cm long, during late
December and January. Found mainly on the dry side of the mountains
of the South Island from the Nelson Mountains southwards, up to 1,400m
in tussock and grasslands.

Family: Compositae

132. Pigmy speargrass, *A. monroi*, in flower, Cupola Basin, 1,700m (December)

133. Plants of false spaniard, Fog Peak, 1,400m (January)

Subalpine spaniard   *Aciphylla pinnatifida*
A small spaniard with leaves up to 20cm long, conspicuously marked by
a deep, yellow, central stripe. The flower stalks are stout and about 15cm
long. Found in high subalpine regions of western Otago and Southland.

134. A plant of the subalpine spaniard in flower above Wapiti Lake, Fiordland (December)

Prostrate coprosmas   *Coprosma pumila, Coprosma petriei*
                                                 and *Coprosma atropurpurea*
Prostrate, creeping and rooting coprosmas often form large, dense, flat
mats amongst grasslands and in herbfields or rocky places in subalpine
and alpine regions throughout New Zealand. The sessile leaves are about
5mm long and 1-3mm wide. Flowers 1.5-2cm high, which occur during
November and December, are erect and conspicuous. The translucent
drupes, 6-8mm long, which are usually ripe by April, persist on the plants
till December or later and may be greenish-white, green, pale blue, orange,
red, purplish-red or claret in colour. (See also page 46.)
                                                 Family: Rubiaceae

35. Prostrate coprosma, *Coprosma pumila*, with red drupes, Cupola Basin, 1,600m (December)

36. Prostrate coprosma, *Coprosma petrei*, with pale drupes, Molesworth (March)

137. Female flowers of prostrate coprosma, *Coprosma atropurpurea*, Key Summit (January)

Brown-stemmed coprosma   *Coprosma acerosa* var. *brunne*

A slender, sprawling plant with interlacing branches that form ope flattened mats up to 2m across in subalpine grasslands, especially alon river terraces and in rocky places from 600m to 2,000m throughout Ne Zealand. The very small flowers, 2-3mm across, occur from August t October, and the translucent drupes, 5-6mm long, mature during Marc and April. North Island plants normally have pale-blue, striped drupe but drupes of many South Island plants are deep, rich blue (fig. 140).

138. Male flowers of brown-stemmed coprosma, Otari (September)

39. Pale striped drupes of brown-stemmed coprosma, Upper Waipunga River (April)

40. Deep blue drupes of brown-stemmed coprosma, Jack's Pass, 900m (March)

Maori onion   *Bulbinella* spp
Maori onion is the popular name for four species of perennial lilies, 30
60cm high with fleshy roots, found in subalpine grasslands and alpin
herbfields from Lake Taupo and Mt Egmont southwards. All hav
conspicuous yellow flowers, which occur from October to January. I
parts of the South Island plants often occur in great numbers over wid
areas, making a brilliant display of colour when in bloom (fig. 145). Flowe
stalks may reach 45cm high, and the flowers, 6-8mm across when full
opened, are borne on the upper third, the lower flowers opening firs
B. hookeri occurs in the North Island and the northern part of the Sout
Island from Nelson south to the Waiau River. *B. angustifolia* is foun
in the South Island east of the Alps, from the Waiau River to Southland
B. gibbsii occurs only on Stewart Island, but *B. gibbsii* var. *balanifer*
is found down the western side of the Alps and is common in the uppe
Hollyford Valley and the Gertrude Cirque. *B. talbotii* is found only o
the Gouland Downs. *B. hookeri* is normally a taller plant than *B. gibbs*
var. *balanifera* and has looser or more open flower heads; its leaves ma
reach 75cm in length, whereas those of *B. gibbsii* seldom exceed 32cm
B. angustifolia leaves reach 100cm in length and tend to have their edge
rolled inwards, especially towards their tips, much more than in eithe
of the other two species. The flower heads of *B. angustifolia* are distinctl
longer than those of either *B. hookeri* or *B. gibbsii*. *B. talbotii* is a prostrat
species found on the Gouland Downs and in the northwest Nelson region
                                                        Family: Liliacea

141. Maori onion, *B. gibbsii* var. *balanifera*, Homer (December)

142. Maori onion, *B. hookeri*, Jack's Pass, 900m
(January)

143. Prostrate Maori onion, *B. talbotii*, a specimen from the Gouland
Downs, grown at Otari by Mr W.B. Brockie (December)

144. Maori onion, *B. angustifolia,* Thomas River Basin (December)

145. Fields of Maori onion, *B. hookeri,* Acheron River Valley, Molesworth (January)

SCREES, ROCKY AND STONY PLACES

VEGETABLE SHEEP

These four extraordinary daisy plants, the giant vegetable sheep (*H. pulvinaris*), the vegetable sheep (*R. eximia*) and the smaller *R. mammillaris* and *R. rubra* all grow in open, windy places on rocks and stones, where they are exposed to great extremes of climate. The places where they are found are often wind funnels, and the surrounding rocks undergo extreme variations in temperature between day and night in the height of summer, while during the winter these plants are often buried under several feet of snow. Their perfect adaptation to such rigorous conditions of living is one of the most remarkable feats of plant evolution. The seeds are windborne and when mature are ejected suddenly in a cluster from the capsule. These seeds were ejected as I watched while taking my photographs and very shortly afterwards were disarranged and moved away by a slight puff of wind.

146. Plant of giant vegetable sheep, *Haastia pulvinaris,* scree above Cupola Basin, 1,650m (April)

147. Surface of *Haastia pulvinaris* showing the densely compacted hairy leaves, Cupola Basin, 1,650m (April)

Giant vegetable sheep   *Haastia pulvinaris*

A rough, hardy, woody shrub, forming large, rounded or flat cushions, usually up to 2m across, but occasionally more. Found throughout the mountains of Nelson and Marlborough, up to 2,750m, in loose rocky places above screes, in fellfields, and on the edges and more stable parts of screes where the rocks are large. The closely packed leaves are covered with long hairs (fig. 147) so that from a distance groups of this extraordinary plant look like flocks of sheep.

The plant in figure 146 was the most magnificent I have ever seen and measured 8m² in area. The flowers of *H. pulvinaris*, which appear mainly during January and February, though flowers can occur as late as April, are small, about 1cm across (fig. 148), and during March and April the masses of dandelion-like seeds are carried away in the wind.

The yellow, forked structures (fig. 148) are the immature stigmas. After pollination they turn dark brown and wither up as the seed matures. A developing seed capsule, almost mature, appears on the right.

Family: Compositae

148. Close view of late flower and developing seed capsules of giant vegetable sheep, Cupola Basin, 1,700m (April)

149. Emerging seeds of giant vegetable sheep, Cupola Basin, 1,700m (April)

Common vegetable sheep   *Raoulia eximia*

Found in dry, rocky places from 1,600 to 2,700m throughout the mountains of Nelson, Marlborough and Canterbury, these plants, like *Haastia pulvinaris*, resemble flocks of sheep on the mountainsides and are adapted for living in extreme climatic conditions. Plants of *R. eximia* form tight, compacted, rounded cushions, 20-60cm high and up to 2m across. The woolly leaves, 3-4mm long by 1-2mm wide, are densely packed and overlapping to form the tight, compact surface of the plant. The small flowers, about 3mm in diameter, occur mostly during January and February though late flowers can persist into April. The downy seeds are adapted for dispersal by wind.

Vegetable sheep grow 'shoulder to shoulder', as in figure 153, right up the sheer rocky slopes of Mt Cupola to the summit at 2,275m, and similar colonies of these plants can be seen on many other mountains in Nelson and Marlborough.

A number of white vegetable sheep like those in figure 152, seen from a distance, look like a small flock of sheep; hence the common name of 'vegetable sheep'.

Family: Compositae

150. Common vegetable sheep, Cupola Basin, 1,700m (April)

51. Late flowers and seeds of common vegetable sheep, Cupola Basin (April)

52. A white form of the common vegetable sheep, Torlesse Range, 1,700m (December).
Photographed on the west face of the leading ridge connecting Mt Plenty to the crest
of the Torlesse Range

153. Common vegetable sheep, showing habit of clothing rock face, Gunsight Pass, Cupola Basin, 1,800m (April)

Red-flowered vegetable sheep   *Raoulia rubra*
A much smaller vegetable sheep, forming tight, hard, rounded, woody
hummocks on rocks at 1,300-1,500m in the Tararua Ranges of the North
Island and the Nelson Mountains of the South Island. Plants are usually
about 15cm high and 25cm across. Leaves, 3-4mm long by 1.5mm wide,
are very hairy, the hairs being longer than the leaves. Red flowers, 2-
3mm across, occur during January and February.

Family: Compositae

154. Red-flowered vegetable sheep, Brown Cow Pass to Boulder Lake, 1,350m (February)

155. Flower of red-flowered vegetable sheep, Brown Cow Pass (February)

Silvery vegetable sheep   *Raoulia mammillaris*
Very similar to *R. eximia* but of a silvery-white colour, with smaller, compacted leaves. Found only in dry, rocky situations on the Torlesse and Craigieburn Mountains. This vegetable sheep forms tough, oval hummocks, up to 50cm across, on windswept slopes and ridges. The tightly packed, overlapping, woolly leaves are 3-4mm long by 1mm wide. Flowers 2mm across occur during late December, January and February, followed by downy seeds.

Family: Compositae

156. Silvery vegetable sheep showing flowers and woolly leaves, Fog Peak, Torlesse Range 1,600m (December)

Haast's buttercup   *Ranunculus haastii*
This unusual-looking buttercup, which grows to 15cm high, is found on the screes in the mountains of Nelson and Canterbury between 1,400 and 1,850m. The deeply divided leaves, up to 10cm wide on petioles 5-15cm long, arise directly from the rootstock, which grows upward through the stones and is fed by long, fine roots that extend down into the damp, sandy zone beneath the sliding shingle. Yellow flowers 2-4cm across occur from November to January. The fruit shown here is a typical buttercup seed head.

Family: Ranunculaceae

57. Haast's buttercup in flower, Gammack Range, near Lake Tekapo, 1600m (December)

58. Haast's buttercup showing divided leaves and seed, scree on Fog Peak, Torlesse Range, 1,500m (January)

159. The tufted haastia in flower, Homer Cirque, 1,250m (January)

Tufted haastia   *Haastia sinclair*

An erect or decumbent, woolly plant, sometimes forming loosely tufte
mats. Leaves, 3.5cm long and 1.5cm wide, are normally covered by
thick, white, partially appressed tomentum, but some plants from Fiordlan
have a buff-coloured tomentum and these are recognised as variety *fulvide*
The flowers, 2-3cm across, occur during December and January. Foun
in subalpine and alpine fellfields, rocky places, and screes from Nelso
to Fiordland, mainly on the east side of the mountains.

Family: Composita

SCABWEEDS OR MAT DAISIES

hese large, tight, flat mats that spread over stones in river beds and
ver rocks in stony and rocky places are popularly known as scabweeds.
lants range in size from quite small to several square metres in area
nd they are seldom more than 2-3cm high. Scabweeds occur throughout
ew Zealand in lowland as well as alpine regions and belong mostly to
ae genus *Raoulia*.

Family: Compositae

.0. Green mat daisy, Bealey Riverbed (January)

reen mat daisy   *Raoulia haastii*
 prostrate, dense, appressed, creeping and rooting herb, forming large,
right-green cushions or hummocks, 1-2m across, covering stones and
aingle in subalpine riverbeds and rocky places east of the Southern Alps
om the Lewis Pass southwards. In this photo the bronze-green winter
•liage is changing to the bright, lush green of the summer season.

Family: Compositae

Open mat daisy in flower *Raoulia glabr*

A prostrate, creeping and rooting herb with lax branches and ascendin
branchlets, forming loose, green patches up to 30cm across. Leaves ar
laxly imbricate, 3-5mm long and 1mm wide. The creamy-white flower
8-10mm across, occur in profusion during December and January. Foun
in lowland and subalpine rocky places, grasslands, herbfields and fellfield
up to 1,300m altitude, from Taupo southwards.

Family: Composita

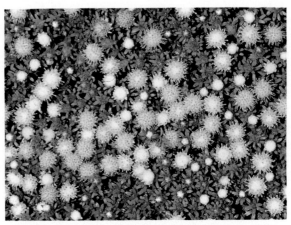

161. Open mat daisy in flower, Lake Lyndon (January)

Scabweed *Raoulia austral*

A close-set, prostrate, creeping and rooting daisy with closely imbricat
leaves, forming greyish mats in subalpine rocky and stony places, ope
ground and fellfields up to 1,600m altitude, from Arthur's Pass to th
Hollyford Valley.

Figure 162 shows the mat-like nature of the plant and the tiny yello
flowers, 3-4mm across, with which it becomes smothered during Januar
and February. Scabweed has a tight, clean edge to the plant.

Family: Composita

Volcanic Plateau raoulia *Raoulia albo-serice*

A greyish-white, spreading and rooting, prostrate, fairly compact her
which forms flat patches up to 60-70cm across. Flowers occur durin
December and January at the tips of erect branchlets bearing closel
imbricate leaves. Found throughout the Volcanic Plateau in bare ston
places, on pumice and in fellfields up to 1,750m.

Family: Composita

162. Scabweed in flower, Bealey Riverbed (January)

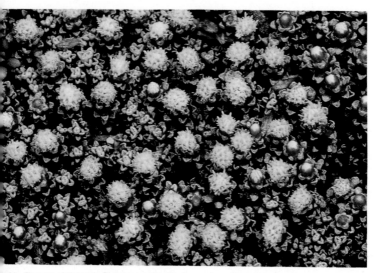

163. Flowers of Volcanic Plateau raoulia, *R. albo-sericea*, Mt Ruapehu (January)

164. Tutahuna plant sprawling over stones, Bealey Riverbed (January)

165. Tutahuna, *Raoulia tenuicaulis*, Outere Stream, Volcanic Plateau (December)

utahuna   *Raoulia tenuicaulis*

creeping and rooting mat daisy with ascending branchlets, forming dense
ats in stony places in river beds and herbfields throughout New Zealand
p to 1,500m altitude. The thick leaves, 5mm long by 2mm wide, taper
om the base to a pointed apex. White flowers, up to 6mm across, occur
uring December and January. Tutahuna has a more open edging around
ie plant, as seen here. The child beside the plant gives some idea of
s size.

Family: Compositae

1ossy scabweed   *Scleranthus uniflorus*

moss-like, dense, perennial herb, forming patches up to 20cm across
i dry riverbeds, grasslands, bare sand, and rocky places throughout the
outh Island up to elevations of 1,250m.

Family: Compositae

6. Mossy scabweed, Bealey Riverbed (January)

167. Flowers and spines of wild irishman, Broken River (January)

Wild irishman   *Discaria toumato*
A common shrub, reaching to 5m high, easily recognised by its tangle
branches bearing many long sharp spines. Found in subalpine riverbeds
grasslands, stony and rocky places throughout New Zealand. The swee
scented flowers occur in profusion from November to January.
Family: Rhamnace

Large white-flowered buttercup   *Ranunculus buchanan*
This lovely plant is found only in damp, rocky clefts between 1,500 an
2,300m, in the mountains from Lake Wakatipu south to around La
Hauroko. The leaves are divided but not hairy, and up to 15cm lon;
The flowers occur in December and January and are 3-7cm across, usual
borne singly though sometimes two or four may occur together.
Family: Ranunculace

Marlborough helichrysum   *Helichrysum coralloid*
Found only in the mountains of Marlborough and distinguished by i
very woolly stems and large, deep-green, appressed leaves, this plant reach
about 60cm in height. It grows in rocky places.
Family: Composit

168. Large white-flowered buttercup above Wapiti Lake, Fiordland (December)

169. Marlborough helichrysum showing woolly stems and appressed leaves (February)

Mountain helichrysum   *Helichrysum intermediu*

Yellow-flowered helichrysum   *Helichrysum parvifoliu*

Two very closely related species of *Helichrysum*, found growing in dr
exposed rocky places up to 1,600m. *H. intermedium* occurs throughor
the South Island, *H. parvifolium* in the mountains of northwest Nelso
Marlborough, and North Canterbury. *H. parvifolium* has larger, yello
flowers, 1cm across, and narrower, more elongate leaves without an acur
cartilaginous tip (fig. 173). *H. intermedium* has smaller, creamy-whit
flowers, 6-7mm across, and broader, more triangular-shaped leaves (fig
171). Flowers occur from December to February or early March. A furthe
species, *H. coralloides*, found only in the mountains of Marlborough, ha
elongate, narrow leaves, stems clothed with long, tangled, white hair
and small flowers like those of *H. intermedium*. A variety of *H. intermediu*
has pointed leaves with acute cartilaginous tips and is known as *I*
intermedium var. *acutum* (fig. 172).

170. Mountain helichrysum, *H. intermedium*, on rocky bluff, Sugarloaf, Cass, 1,450n
(November)

171. Close-up of stem with appressed leaves of *H. intermedium*, Cupola Basin (December)

172. Close-up of stem with appressed leaves and acute tips of *H. intermedium* var. *acutum*, Jack's Pass (January)

173. Flowers of yellow-flowered helichrysum, *H. parvifolium*, Jack's Pass (March)

174. Flowers of leafless clematis, Marlborough (October)

Leafless clematis   *Clematis afolia*
A vine forming masses of tangled stems up to 1m high, found growing
on open rocky places to elevations of 1,000m from Hawke's Bay to
Southland. Flowers 3-4cm across occur profusely during October and
November, and the seeds set rapidly as the flowers die off.

Family: Ranunculaceae

175. Seeds of leafless clematis, Marlborough (November)

176. Flowers of porcupine plant, Cupola Basin (December)

177. Berries of porcupine plant, Cupola Basin (February)

Porcupine plant   *Hymenanthera alpina*

A stiff, spreading shrub with rigid, interlacing branches, each bearing many lenticels and usually terminated by a stout spine. The plant reaches 60cm in height and may be 1m across. The leaves are 6-18mm long, and the tiny bell-shaped flowers hang in thousands along the branches during November and December. The white berries, 5mm across, are ripe in February. Found growing in exposed rocky places between 600 and 1,300m altitude, east of the Southern Alps.

Family: Violaceae

178. Leafless porcupine plant in flower, Tararuas (November)

Leafless porcupine plant   *Hymenanthera angustifoli*
A rigid, springy shrub, up to 3m high, similar in habit to *H. alpina* but
usually without any leaves. The minute flowers, 2-3mm across, occur i
thousands along the branches during October and November, and produc
clouds of pollen, which float out of the bush with the slightest breeze
The white, purple-spotted berry, 3-4mm in diameter, is mature durin
November or December. Found in lower subalpine, stony riverbeds an
along forest margins throughout New Zealand.

Family: Violacea

Barbless bidibidi   *Acaena glabr*
A creeping, hairless plant with prostrate rooting branches rising at thei
tips and up to 50cm long, with leaves up to 5cm long; the flower heads
up to 2cm in diameter, either green- or red-coloured, occur from Decembe
to February. The plant is found on screes and in subalpine riverbed
throughout the eastern South Island.

Family: Rosacea

South Island edelweiss   *Leucogenes grandicep*
A herb, woody at the base with ascending branches bearing loosel
overlapping leaves 5-10mm long by 2-4mm wide and densely clothed wit
silvery-white wool. The flower head, 9-15mm across, is surrounded by
about 15 very woolly leaves, each up to 1cm long, which form a distinc
circlet of white rays. Flowers occur from November to February, and
the plant is found throughout the South Island in rocky places, rock
ledges and moraines up to 1,600m.

Family: Composita

79. Barbless bidibidi in flower, Mt Torlesse Range (December)

80. South Island edelweiss in flower, Homer Cirque, 900m (January)

181. Glossy willow-herb with flowers and buds,
Gertrude Cirque (January)

Glossy willow-herb   *Epilobium glabellum*
A very variable willow-herb, forming patches up to 70cm across and found
commonly in mountain regions from East Cape southwards up to 1,800m,
in fellfields, stony riverbeds, and among rocks or grasslands. The elongate
flower buds are characteristic of willow-herbs.

Family: Onagraceae

Red-leaved willow-herb   *Epilobium crassum*
Shown here nestling in front of a *Haastia pulvinaris* plant on a scree
above Cupola Basin, this plant shows the characteristic curled seed capsules
of willow-herbs. This *Epilobium* is woody at its base, with stems 7-15cm
long and small, densely crowded leaves 6-8mm long. Flowers 5-7mm across
occur from December to February. Found in rocky places and on screes
amongst the mountains from Nelson to Arthur's Pass between 900 and
1,800m.

Family: Onagraceae

182. Large flowering plant of glossy willow-herb, 60cm across, Gertrude Cirque (January)

183. Red-leaved willow-herb in seed, Cupola Basin (April)

Fern-leaf pincushion   *Cotula squalida*
A prostrate, branching, spreading and rooting herb with wiry branches
forming patches up to 50cm across. Branchlets and leaves are clothed
with slender, silky hairs. The leaves, 2.5-5cm long, are divided into 8-
12 pairs of broad pinnae, giving a decidedly fernlike appearance. Flowers
5-6mm across on stalks 5-20mm long, occur from November to February.
Found in grasslands, exposed stony places and damp or shaded places
up to 1,400m, from East Cape to Stewart Island.

Family: Compositae

184. Fern-leaf pincushion in flower, Ruahine Mountains (December)

Mountain cress   *Notothlaspi australe*
A fleshy herb with somewhat pubescent leaves, 1-5cm long, forming rosettes
up to 12cm across. During December each rosette becomes smothered
by many strongly fragrant flowers, each up to 1cm across. Mountain cress
flowers when quite small and in its first year of growth. Plants apparently
live for several years, adding at least one rosette of leaves each year from
the deep-running rootstock. Found on alpine screes, where the stones are
small and somewhat stabilised, above 1,600m and mostly between 1,600
and 1,800m among the mountains of Nelson and Marlborough. Many
plants often occur together in restricted areas, usually on screes facing
between north and west.

Family: Cruciferae

85. Mountain cress, on scree above Cupola Basin, 1,700m (April)

86. Mountain cress in flower, scree above Cupola Basin, 1,600m (December)

187. Black daisy in flower, scree on.Fog Peak, Torlesse Range, 1,500m (January)

188. Penwiper plant, on a scree on Sugarloaf, Cass, 1,300m (December)

Black daisy   *Cotula atrata*

A creeping, scree-inhabiting plant with down-covered stems up to 20cm long, ascending at their tips and bearing clusters of thick, fleshy, downy, pinnatifid leaves up to 20mm long. Flower heads, up to 2cm across, are borne on short, stout stalks during January and February and may be either black or dark brown. Found on alpine screes between 1,250 and 2,000m altitude among the mountains of Marlborough and Canterbury.

Family: Compositae

Penwiper plant   *Notothlaspi rosulatum*

A fleshy herb up to 25cm high when in fruit, and forming rosettes up to 8cm across; growing on screes amongst finer debris that is partially stabilised. Found on the east side of the Southern Alps from Marlborough to South Canterbury, at 800-1,850m altitude. The leaves, up to 2cm long on petioles 2cm long, are overlapping with serrate or dentate margins. The highly fragrant flowers, similar to those of *N. australe*, occur during December and January on a stout stalk. Penwiper is a transient plant, always difficult to find and seldom occurring in large numbers in any one area for more than two or three years. It probably takes two years to mature, after which it flowers, seeds and dies. Very small plants, sometimes found with flowers, may arise from the stalks of older plants sheared off by moving stones. Buds open from the base of the stalk upwards, and the seeds form around the base while the apical flowers are open.

Family: Cruciferae

89. Small penwiper plant in flower and showing seeds set at the base of flower stalk, scree, Porter's Pass, 1,250m (January)

190. Bristly carrot in flower, Temple Basin, Arthur's Pass (January)

Bristly carrot   *Anisotome pilifera*
A robust plant up to 60cm tall, with pinnate leaves 10-30cm long by 5-
10cm wide on stout petioles about 10cm long. Flowers occur as compound
umbels from November to March. Found in the South Island mountains
in rocky places from northwest Nelson southwards, up to 1,850m.
Family: Umbelliferae

Kopoti   *Anisotome aromatica*
Kopoti and the 14 allied species of *Anisotome* found in New Zealand
are common plants in rocky, alpine places throughout the country. Kopoti
is very variable in form, and several distinct varieties are recognised. At
high altitudes the plants are dwarfs, only 10cm high, but at lower level
plants reach 30-50cm in height. The aromatically scented flowers occur
as umbels from October till February. The species and its varieties are
distributed throughout New Zealand, growing in alpine and subalpine rocky
places, grasslands, herbfields and fellfields.
Family: Umbelliferae

191. Dwarf form of kopoti, Mt Ruapehu, 1,600m
(December)

192. Kopoti in flower, Mt Ruapehu (January)

Haast's carrot   *Anisotome haast*
An *Anisotome* up to 60cm tall with stout, grooved, purple-coloured stem
bearing two- to three-pinnate, rather carrotlike leaves, 15-25cm long an
6-12cm wide, on stout petioles 8-12cm long. Flowers occur from Octobe
to February. Found throughout the South Island in rocky places an
fellfields, mainly on the wet western side of the Southern Alps betwee
500 and 1,600m.

Family: Umbellifera

193. Haast's carrot in flower, Arthur's Pass, 1,200m
(January)

Creeping lawyer   *Rubus parvu*
A prostrate, creeping and rooting bramble with a red bark and serrate
leaves 2-9cm long, found in subalpine, dry, stony places and along ston
banks of river valleys in Nelson and Westland up to 900m altitude. Th
flowers, about 2cm across, occur singly from September till Decembe
and the fruit, up to 2.5cm long, matures from February to April.

Family: Rosacea

194. Creeping lawyer in flower, Boulder Lake
(November)

195. Creeping lawyer in fruit, Boulder Lake (February)

196. Mountain sandalwood with fruits, Boulder Lake (March)

Mountain sandalwood   *Exocarpus bidwil*
A stiffly branched, rigid shrub up to 60cm high. Parasitic on roots, wi
its leaves reduced to minute, triangular scales 0.5mm long. It is four
in exposed, alpine and subalpine, dry, sunny, rocky places from Nels
to North Otago. Minute flowers occur from December to February ar
the nuts, 5mm long in their red arils, take 12-14 months to mature.
                                        Family: Santalace

MOUNTAIN HEBES

lants of the genus *Hebe* are common in rocky places throughout the
ountains of New Zealand, and a selection of these alpine species is shown
ere. All belong to the family Scrophulariaceae.

7. Flowers of black-barked mountain hebe, Cupola Basin (January)

lack-barked mountain hebe   *Hebe decumbens*
spreading, decumbent shrub with shining, purplish-black or brown bark
nd spreading, concave, red-edged leaves 1-2cm long. The flowers occur
om November to February, and the plant is found on rocky ledges and
ony places from Nelson to Canterbury, up to 1,400m.

198. Flowering plant of spiny whipcord, Cūpola Basin (December)

199. Flowers of spiny whipcord, Porter's Pass (November)

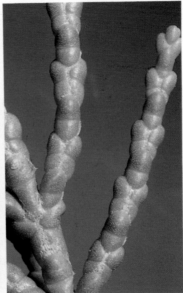

200. Stem structure of spiny whipcord (November)

201. Close view of stem of cypress whipcord, Mt Maungatua (December)

Cypress whipcord   *Hebe propinqua*

A much-branched, usually erect but sometimes decumbent shrub up to 1m tall but generally of less height and forming a dense, flat-topped cushion or mat about 1m across. The appressed, rounded leaves, 1-1.5mm long, are joined together along their basal half. The very small flowers, 5-6mm across, occur from December to February as spikes of 12, each about 1cm long, forming small terminal heads to the branchlets.

Found in the mountains of South Canterbury and Otago between 800 and 1,500m, in damp peaty ground amongst rocks in scrub and fellfields.

Spiny whipcord   *Hebe ciliolata*

A stiff, whipcord *Hebe*, up to 30cm tall, found in rocky places among the mountains of Nelson and Canterbury up to elevations of 2,400m. The leafy twigs, 2.5-7mm in diameter, bear appressed leaves with ciliate or spiny margins as shown in figure 200. The numerous stomata are also clearly shown in this photo. The terminal flowers are sessile, arranged as one to three pairs, and occur from November till January (fig. 199).

Nelson mountain hebe   *Hebe gibbsii*
Found on the Nelson Mountains, Ben Nevis and Mt Rintoul, this sparingly
branched shrub reaches a height of 30cm. One of the imbricate-leaved
hebes, it is readily recognised by the red edging to the thick leaves, 10-
18mm long by 6-12mm wide, and by the fringe of hairs round each leaf
margin. The sessile flowers occur on terminal spikes 1.5cm long from
December to February.

202. Flowers of Nelson mountain hebe, Dun Mountain (December)

Dish-leaved hebe   *Hebe treadwellii*
An erect shrub up to 15cm high, with grey-green foliage and black stems,
found commonly from Nelson to South Canterbury, amongst subalpine
or alpine scrub and fellfields, in dry areas of low rainfall. The thick, shining,
broadly based green leaves, up to 25mm long by 12mm wide, are imbricating
or spreading, and the crowded white flowers, each with a short pedicel,
occur from October till April. A similar plant, but with shorter, 5-10mm
long, glaucous leaves and sessile flowers, is known as *H. pinguifolia*.

Takaka hebe   *Hebe divaricata*
An erect, branching shrub up to 3m tall, found along streamsides and
in rocky places throughout Marlborough and Nelson up to 1,100m altitude.
Shown here growing to advantage amongst rocks, this species illustrates
the fine, showy hebes of New Zealand. The branchlets are slender and
very finely pubescent, with wide spreading, narrow, lanceolate leaves 2-
3cm long by 3-4mm wide. Flowers in lateral racemes occur profusely during
January and February. This species is confined to Marlborough and Nelson,
where, however, it is relatively common.

203. Dish-leaved hebe with flowers, Cass
(December)

204. Takaka hebe in full flower, summit Takaka Hill (January)

205. Colenso's hebe in full flower, Taruarau River (October)

206. Flower racemes and leaves of Colenso's hebe (October)

207. Flowers of New Zealand lilac, Marlborough (November)

New Zealand lilac   *Hebe hulkeana*

Found in its natural state in dry rocky places over the northern section of the South Island up to 900m, this rather loosely branched shrub reaches a height of 60cm. The branchlets are covered with fine hairs, and the spreading, elliptic to obtuse-shaped leaves, 7-10cm long by 2-3cm wide, have serrated or dentate margins. Flowers occur in profusion as crowded terminal spikes from October till December.

Colenso's hebe   *Hebe colensoi*

A somewhat spreading hebe of low-growing, open habit, found around the headwaters of the Taruarau River and on the Ruahine Mountains from the Kaweka Range south to the Moawhango River. This species grows on bank or cliff faces overlooking river valleys and is characterised by its smooth, thick, blue-green leaves, 2-5cm long and 1cm wide. The flowers occur from August to November as lateral racemes near the tips of the branches, each raceme slightly longer than the leaves.

209. Canterbury whipcord, *H. Cheesmanii*, plant in full flower, 1,350m, off Cass River above L. Tekapo (December)

Canterbury whipcords   *Hebe cheesmanii, Hebe tetrasticha*

Shrubs 20-30cm high with erect, leafy stems 1.5-2.5mm in diameter, stouter in *tetrasticha*. Leaves crowded and appressed in *tetrasticha* to branchlets 1.5mm long by 1mm wide with narrow tips; in *cheesmanii,* 1mm long by 1.5mm wide, each with a faint keel towards an acute tip. The sessile flowers in one to three pairs in *tetrasticha* and usually in two pairs in *cheesmanii.* If the twigs of the species are examined in cross-section, it will be seen that the twigs of *H. testraticha* are more 'cross' shaped, while *H. cheesmanii* has more 'square' shaped twigs. Both species are found among the mountains of Canterbury up to an altitude of 1,800m.

◄208. Canterbury whipcord, *H. cheesmanii,* showing hairy leaf margins and flowers, above Cass River, Godley Valley (December)

Large-flowered hebe   *Hebe macrantha*
Usually a rather untidy-looking shrub about 60cm high, with erect branches
bearing thick, leathery leaves 1.5-3cm long, having serrated or toothed
margins. Flowers about 18mm across occur at the tips of the branches
as racemes of three to eight flowers each, during December, January and
February. The plant is found on steep, rocky faces, often in alpine scrub,
among the South Island mountains between 800 and 1,600m.

210. Flowers of large-flowered hebe, Wilberforce River Valley
(December)

Blue-flowered mountain hebe   *Hebe pimeleoides*
An erect, branching hebe, not more than 45cm high, with narrow,
lanceolate, bluish-coloured leaves, 5-15mm long by 2-6mm wide, found
in the drier rocky places of the mountains from Marlborough southwards.
Flowers, arranged in pairs on lateral, hairy stalks, from November to
March, vary from bluish-white to deep blue in colour.

Leathery-leaved mountain hebe   *Hebe odora*
A common shrub up to 1.5m high, found in damp, stony places amongst
herbfields and fellfields in the mountains of New Zealand between 800
and 1,600m. The thick, leathery, concave leaves, 2-3cm long, have a distinct
keel, faintly crenulate margins and stout petioles, which are twisted to
bring the leaves into two opposite rows.
   Flowers occur as lateral and terminal spikes from October to March.
In Fiordland plants the leaves are longer, narrower, and more distinctly
keeled.

211 Blue-flowered mountain hebe with flowers,
     Upper Rangitata (December)

212. Leathery-leaved mountain hebe, Fiordland form, in flower, Gertrude
      Cirque, Homer (January)

213. Leathery-leaved mountain hebe, Mt Oxford (December)

Trailing whipcord   *Hebe haasti*

A prostrate or sprawling shrub with decumbent, woody stems, ascending at their tips and up to 40cm long. The leathery, rigid, recurved leaves 6–13mm long by 4–9mm wide, are arranged in four overlapping rows, their partially hairy leaf margins bordered but not thickened. The terminal flowers, about 8mm across, occur from November to February. Found or screes along the dry side of the South Island mountains between 900 and 2,600m. The very similar *H. epacridea* occurs in similar situations but has thickened leaf margins.

214. Close view of trailing whipcord with flowers, Mt Terako (December) ▶

215. Mt Arthur hebe, Mt Arthur, Nelson (December)

216. Close view of spray of Mt Arthur hebe in flower,
Mt Arthur (December)

117. Flowers of pink hebe, Mt Terako (November)

Pink hebe *Hebe raoulii*

A straggling, sprawling shrub with branches up to 30cm long and obovate leaves 6-25mm long by 2.5-8mm wide. Flowers occur abundantly from October to March in terminal racemes 2-6cm long. Found in dry, rocky places amongst the mountains of Nelson, Marlborough and Canterbury up to 900m.

Mt Arthur hebe *Hebe albicans*

A closely branched, low-spreading shrub up to 1m high, with brown bark. Imbricate, rather elongate, sessile, bluntly pointed, thick, bluish-coloured leaves are 1.5-3cm long by 8-15mm wide. Flower heads 3-6cm long occur from December to April. Found in stony places amongst the mountains of Nelson from 1,000 to 1,400m. A closely related species, *H. amplexicaulis,* has more obtuse-tipped leaves and is found in similar situations amongst the mountains of Canterbury.

218. Cushion plant, Fog Peak, Torlesse Range, 1,750m (January)

Cushion plant   *Pygmea pulvinari*
A soft, moss-like herb forming a compact, dense cushion, 2-4cm high
and about 10cm across, in subalpine and alpine, rocky and shingly place
of the South Island from Nelson to Canterbury. The tiny, linear-spathulate
or linear-oblong leaves, 2.5-4mm long and 1mm wide, are sparingly clothed
with long, bristle-like hairs. Flowers 5-6mm across appear during Decembe
and January.

Family: Scrophulariaceae

Snowball spaniard   *Aciphylla congest*
A soft-leaved *Aciphylla* forming compact masses of rosettes and foun
growing beside water trickles and alongside damp cracks in rocky outcrop
and basins in the mountains of West Otago and Fiordland, between 1,30
and 1,600m. Leaves are thin and flexible, forming open mats up to 60cm
across. They are up to 6cm long, tapering in width from 2cm to 7mm
Flowers occur during January as large, white, rounded umbels up to 12cm
diameter, which, in the distance, look like snowballs lying on the rocks.

Family: Umbelliferae

Prostrate dwarf broom   *Carmichaelia enys*
A low shrub not more than 5cm high, which forms patches up to 10cm
across and which is leafless in the adult stage. Flowers, 5mm across an
4mm long, occur as one- to three-flowered racemes on stalks 5mm lon
from November to January. Found on dry, stony river terraces an
grasslands up to 1,400m, along the eastern side of the Southern Alps from
Arthur's Pass to Fiordland.

Family: Papilionaceae

19. Snowball spaniard, Gertrude Cirque, 1,400m (January)

20. Prostrate dwarf broom in flower (December)

Rock cushion   *Phyllachne colenso*

A plant forming compact, hard cushions or mats, up to 40cm across
among rocks in herbfields, fellfields and rocky places up to 1,850m. The
leaves, 4mm long, are thick and leathery. The small, white flowers that
appear during January and February have their anthers extended far beyond
the lip of the corolla. Found from Mt Hikurangi southwards to Stewart
Island.

Family: Stylidiaceae

Fleshy lobelia   *Lobelia roughi*

A small, spreading, hairless herb, 5-10cm high, found on shingle scree
and rocky outcrops from the Nelson Mountains to Otago, between 900
and 1,850m. The stems and branches twist between the stones, and the
tough branchlets arise erect above the stones, bearing thick, leathery leaves
10-20mm long by 10-15mm wide, on broad, flat petioles 5mm long. Flowers
10mm long occur on stout stalks up to 5cm long from October to February.
Seed capsules 10-12mm long mature from December till April. The cut
stems yield a very bitter-tasting, yellow sap.

Family: Lobeliaceae

221. Fleshy lobelia in flower, Torlesse Range (January)

22. A plant of rock cushion fits snugly between stones at 1,600m on Mt Lucretia (January)

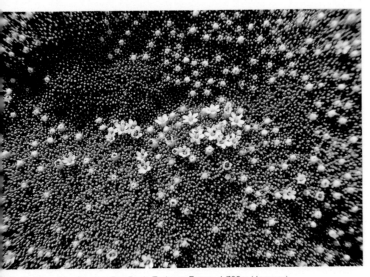

223. Rock cushion flowers, Fog Peak, Torlesse Range, 1,700m (January)

224. Snowy woollyhead in flower, scree on Porter's Pass, Torlesse Range, 1,350m (January

225. Mountain chickweed in flower, scree on Fog Peak, Torlesse Range, 1,400m (January)

Snowy woollyhead *Craspedia incana*

A soft plant, covered all over with dense, silvery-white wool, found among rocks and screes of the dry mountains of Canterbury. The leaves, 5-10cm long by 2-3cm wide, are arranged as rosettes, and the flower heads, 2-3cm across, occur during January and February.

Family: Compositae

Mountain chickweed *Stellaria roughii*

A branching, succulent, perennial herb with leaves 20mm long by 3mm wide, found on the screes of the South Island mountains between 1,000 and 2,000m. The flowers have sepals longer than the petals, and the sepalate flower may reach 2cm across with the corolla of petals only 1cm across.

Family: Caryophyllaceae

Grey-leaved succulent *Lignocarpa carnosula*

Until recently regarded as an *Anisotome,* this plant with thick, fleshy stems has been placed in a new genus, *Lignocarpa.* It forms almost succulent plants up to 15cm tall, found among loose stones on screes of the Torlesse Range and other mountains in Canterbury and Marlborough from 1,200m to about 1,600m. The tiny flowers, 3-4mm across, occur from November to February.

Family: Umbelliferae

226. Grey-leaved succulent in flower, Fog Peak, Torlesse Range, 1,400m (January)

227. Scree epilobium, Fog Peak, Torlesse Range, 1,700m (January)

Scree epilobium   *Epilobium pycnostachyun*
A low, spreading plant, arising from a woody stock, found commonly
on screes between 1,200 and 1,800m from the Ruahine Mountain
southwards to the Humboldt Range. Stems up to 25cm long bear green
or deep-red leaves, 10-20mm long and 2-4mm wide, on short petioles 2mm
long. Small, white, star-like flowers occur at the tips of the stems from
October in the North Island and from December in the South Island until
March.

Family: Onagraceae

Succulent daisy   *Vittadinia australi*
A small, bushy, fleshy, low-growing plant, 10-30cm across, found growing
amongst rocks, on screes, and in rocky places in herbfields and grassland
up to 1,100m throughout New Zealand. The flowers, about 10mm across,
occur from November to March, and the deeply toothed leaves are up
to 1.5cm long.

Family: Compositae

Bronze crane's-bill   *Geranium sessiliflorum* var. *glabrum*
A perennial herb with very short stems, usually forming small clumps.
Leaves may be light or dark green or bronze-coloured on the same plant,
10-20mm across, very hairy, with petioles usually about 5cm long. Flowers
1-2cm across occur from December to April. Found often on screes but
also in stony places among grasslands and herbfields up to 1,100m
throughout New Zealand.

Family: Geraniaceae.

228. Succulent daisy in flower, scree above Porter's Pass, 1,000m (January)

229. Plant of bronze crane's-bill in flower, Lake Pukaki (January)

230. Close-up of bronze crane's-bill in flower, scree above Porter's Pass, 1,000m (January)

231. Flowers of small forget-me-not, Dun Mountain (November)

232. Arthur's Pass forget-me-not, Arthur's Pass (December)

Arthur's Pass forget-me-not   *Myosotis explanata*

A perennial, rosette-shaped herb up to 30cm high with leaves 3-7cm long covered all over by soft, white hairs. Flowers, 12-16mm across, occur from December to February as terminal cymes. Found in rocky places between 900 and 1,400m in the vicinity of Arthur's Pass.

Family: Boraginaceae

Small forget-me-not   *Myosotis monroi*

A prostrate, rosette-forming plant with hairy leaves. Flowers are produced in cymes during October, and the plant is found in rocky places on Dun Mountain, Nelson.

Family: Boraginaceae

Colenso's forget-me-not   *Myosotis colenso*

A prostrate, creeping perennial, spreading by decumbent branches up to 6cm long. The lanceolate leaves, 2-3cm long by 5-10mm wide, have more hairs on the upper surface than on the lower. Flowers 8mm across may be single or in clusters during November and December. Found in Canterbury on limestone rocks around Broken River.

Family: Boraginaceae

233. Colenso's forget-me-not, Castle Hill (November)

Rock daisy   *Pachystegia insigni*

A strong, spreading shrub, occasionally reaching 2m high but usually much less, found on cliffs and rocky places throughout Marlborough from sea level up to 1,250m altitude. The thick leaves, 6.5-16cm long, are on petiole 5cm long and are clothed below with dense, soft hairs. The flower bud bear a characteristic scale pattern, and flowers up to 7.5cm across occur from December to February.

Family: Compositae

Northern snowberry   *Gaultheria colense*

A low-growing, sprawling or bushy shrub up to 60cm high with thick leaves, either crenulate or entire, 8-12mm long by 5-9mm wide, on petiole 1mm long. Flowers 2-3mm long occur in profusion from November to January, and the berries, 3mm in diameter, ripen during February and March. Found in rocky places and grasslands amongst the North Island mountains from Mt Hikurangi southwards.

Family: Ericaceae

234. Rock daisy in flower, Kaikoura Mountains (January)

235. Northern snowberry flowers, Mt Ruapehu (January)

236. Maruru flowers, Mt Egmont (December)

Maruru   *Ranunculus hirtu*

A slender, branching ranunculus, bearing flowers 1.5cm across on lon,
hairy stems up to 60cm high from September to February. The hair
bipinnate leaves, 2-5cm long, are on slender petioles about 5cm long. Th
ranunculus is found all over New Zealand. In subalpine regions it occui
commonly in scrub or in rocky places.

Family: Ranunculacea

BOGS AND SWAMPS

gs and swamps may be extensive in area or very small, occurring near
pages at the bases of slopes and rocky bluffs in the mountains. The
gs on Key Summit are among the most extensive, but small bogs and
amps abound everywhere in the New Zealand mountains.

7. Alpine sundew in flower, with *Aporostylis bifolia* on the right and *Donatia* behind, Key
Summit (January)

lpine sundew   *Drosera arcturi*
n insectivorous plant found in mountain bogs and swamps from the
olcanic Plateau to Stewart Island. The narrow, straplike, non-spathulate
aves with their petioles are 5-12cm long, they bear sticky hairs that,
in all sundews, entangle insects and thus the plant obtains its nitrogen.
owers up to 16mm across on stalks 15cm high occur from November
March.

Family: Droseraceae

238. Wahu, Key Summit (January)

Wahu   *Drosera stenopeta*
A fairly common sundew found in lowland and alpine bogs up to 1,600
altitude. The narrow, strap-like leaves, expanded and spathulate at the
tips, are 2cm long with petioles 5cm long but without stipules. The flowe
7-10mm across, occur from November to March. A rather similar sunde
D. spathulata, has leaves 5mm long on petioles up to 10mm long wi
stipules up to 7mm long.

Family: Droserace

Scented sundew   *Drosera bina*
Found in lowland to subalpine bogs and swamps all over New Zealan
This unusual sundew has each leaf divided into two or more narrow forl
arising from the erect petiole and bearing glistening gland hairs. The petio
may be 35cm long and the leaf 15cm, giving a total height of 50cm. Blac
stalks up to 50cm high bear corymbs of large, white, sweet-scented flower
each up to 2cm across, from November to February.

Family: Droserace

39. Wahu in flower, Lewis Pass (December) (see also fig. 437)

40. Scented sundew, Te Anau Bogs (January)

241. Alpine cushion in flower, Arthur's Pass (January)

Alpine cushion   *Donatia novae-zelandia*
A small, firm-tufted plant, which forms hard, broad cushions in alpine
bogs and damp herbfields from the Tararua Ranges southwards. Sessile
flowers, 8-10mm across, occur in profusion during January and February.
(See also figs. 435 and 436.)

Family: Donatiaceae

242. Common nertera with drupes, Otira Gorge (April)

243. Horizontal orchid, Key Summit (January)

Horizontal orchid *Lyperanthus antarcticus*
An unusual orchid in that the flower lies horizontally, almost at right
angles to the stem. The leaves are 2.5-7cm long, and the flowers 8-14mm
long. Found in subalpine bogs from the Tararua Ranges to Stewart Island.
Family: Orchidaceae

Common nertera *Nertera depressa*
A prostrate, creeping plant forming patches up to 30cm across in damp,
shady places. More commonly a plant of lowland forests and streamsides,
this nertera does, however, occur quite frequently in subalpine regions
on damp, shady banks, in damp scrub and bogs in herbfields. Minute
flowers occur from November to February, and the drupes ripen from
January to May.
Family: Rubiaceae

244. Creeping lily in flower, Wilderness (January)

245. Streamside portulaca in flower, Wapiti Lake, Fiordland, 900m (December)

246. Moss daisy in flower, near Boulder Lake (February)

Moss daisy   *Abrotanella caespitosa*

Eight species of *Abrotanella* occur in New Zealand and all are more or less moss-like in general appearance. *A. caespitosa* has thick, leathery leaves, -1.5cm long by 1-1.5mm wide, forming rosulate, matted patches up to 0cm across. The minute flowers occur during January and February, and the plant is found on the edges of bogs and seepages in subalpine grasslands and herbfields from the Ruahine Ranges to Northern Fiordland.

Family: Compositae

Creeping lily   *Herpolirion novae-zelandiae*

A creeping, wiry plant with grass-like leaves forming patches up to 1m across in damp places and along the edges of bogs, in lowlands and mountains up to 1,250m. Flowers 2-3cm across, which may be white, pale blue or pale mauve, occur during January and February. This plant is one of the smallest lilies found anywhere in the world.

Family: Liliaceae

Streamside portulaca   *Montia australasica*

A succulent, creeping and rooting, perennial herb that forms dense or loose patches up to 30cm across along subalpine streamsides where spray or seepages keep it damp, or in wet boggy grasslands and herbfields, often associated with mosses. The long, narrow leaves, usually from 1-5cm long, are on slender petioles, and flowers up to 2cm across, which may be pink as well as white, occur singly, in pairs or as small racemes from November to January. Found from Mt Egmont and the Volcanic Plateau southward to Stewart Island.

Family: Portulacaceae

247. Prostrate grass tree in flower, Key Summit (January)

Prostrate grass tree   *Dracophyllum muscoide*
A stout-stemmed, creeping grass tree with brownish-black stems up t
30cm long. Branchlets, about 4cm long, are densely clothed with thick
leathery, imbricating leaves up to 6mm long and 2.5mm wide. Singl
terminal flowers 4mm long occur during January and February. Foun
around subalpine bogs and in damp grasslands or herbfields of the Sout
Island from Lake Ohau southwards.

Family: Epacridacea

Niggerhead   *Carex sect*
A large sedge up to 1.5m high, whose matted roots and decaying leave
often form huge, broad pillars 1-1.5m high, which lift the sedge and form
the conspicuous objects called 'niggerheads' in swamps. The leaves ar
1-2m long and 3-4mm wide, keeled below, flat above, with sharp cuttin
(scabrid) margins. Flower panicles, which can reach 1m in length, occu
from December to February. *C. secta* is found throughout New Zealan
along stream banks, in swamps and bogs, and in other damp places i
both subalpine and lowland regions.

Family: Cyperacea

248. Niggerheads in swamp, Thomas River Valley
(January)

249. Niggerhead in flower, Volcanic Plateau (December)

Small alpine swamp sedge   *Carex sinclairii*

A short-leaved sedge found in swamps amongst the North Island mountains between 500 and 1,200m but occurring from sea level to 1,300m in the South Island. Tufted flowers occur from December to February.

Family: Cyperaceae

250. Small alpine swamp sedge, Key Summit (January)

Native oxalis   *Oxalis lactea*

Generally regarded as a plant of lowland damp and shaded places, the native oxalis also occurs in subalpine and alpine regions, and I have seen it in a bog in Cupola Basin at 1,300m altitude. Spreading by creeping rhizomes, *O. lactea* produces tufts of three-lobed leaves at intervals, forming patches up to 20cm across. The funnel-shaped flowers, 1.5-2cm in diameter, occur from October to March. Found around bogs and in wet places along stream banks throughout New Zealand, it is one of the three New Zealand species of oxalis.

Family: Oxalidaceae

251. Turf-forming astelia with fruit, Fiordland (December)

Turf-forming astelia   *Astelia linearis*

Two low-growing, turf-forming astelias, *A. linearis* and *A. subulata,* occur
n wet or boggy places among the South Island mountains from the Paparoa
Range to the Longwoods. This plant, photographed in the Upper Stillwater,
Fiordland, at about 1,400m, appears to be a form of *A. linearis.* The
eaves are keeled and about 10cm long.

Family: Liliaceae

252. Native oxalis flowers, Upper Travers Valley (December)

253. Mountain bog daisy, Lewis Pass (January)

Mountain bog daisy   *Celmisia alpine*
A small daisy found commonly in bogs among the South Island mountains
between 600 and 1,600m. The sharp-pointed leaves, 15-30mm long, which
arise as tufts at the tips of short branches, are mottled with brown or
white and have their margins rolled inwards almost to the midribs. A
single flower stalk 5-8cm long arises from each leaf tuft to carry a flower
1.5-2cm across. Flowers occur in profusion from December to February.
                                              Family: Compositae

STREAMSIDES, DAMP OR SHADED PLACES

54. Dense nertera with drupes, Lake Tekapo (February)

Dense nertera *Nertera balfouriana*

A compact, creeping plant forming dense mats up to 25cm across on damp ground and in sphagnum bogs between 600 and 1,000m. Both leaves and flowers are 2-3mm long, and the drupes, 7-9mm long, ripen during February and March. *Nertera* species are found all over New Zealand, forming patches up to 40cm across. Flowers and drupes occur throughout much of the year in most species and they grow best in subalpine, damp, shady places but can tolerate drier conditions.

Family: Rubiaceae

255. Ciliate nertera with flowers and ripe drupes, Mt Egmont (November)

256. Ciliate nertera flowers and leaves showing cilia, Mt Egmont (November)

257. Swamp musk in flower, Upper Travers Valley (December)

Swamp musk *Mazus radicans*

A creeping and rooting perennial herb with short, erect, leafy branches, 2-5cm long, found in damp places and along edges of bogs in subalpine regions up to 1,200m, from Mt Egmont and the Huiarau Ranges southwards to Fiordland. The leaves and petioles, 2-5cm long, are hairy, and the flowers, 1.5-2cm across, occur from November to March.

Family: Scrophulariaceae

Ciliate nertera *Nertera ciliata*

This nertera is distinguished from the common nertera (fig. 242) by the short hairs that occur sparingly on the stems and leaf margins. Leaves are 3-5mm long on petioles 1-3mm long; flowers are about 2mm across, and the drupes are 4mm across.

Family: Rubiaceae

False buttercup   *Schizeilema haastii*
Plants of the genus *Schizeilema* look like ranunculi at first sight, but close
examination shows the stout leaf petioles and the tiny umbelliferous flowers.
S. haastii is found growing along cracks in steep rock faces where water
seeps out, and among rocks in fellfields between 1,000 and 1,400m. The
leaves are 2-3cm across on petioles 3-7cm long, and the tiny purplish flowers
occur as umbels 15mm across during December and January.

Family: Umbelliferae

Mountain pincushion   *Cotula pyrethrifolia*
A strongly aromatic herb with much-branched, creeping and rooting stems
bearing deeply pinnatifid leaves. Found throughout the New Zealand
mountains in damp grasslands and herbfields, along streamsides and in
damp, stony places up to 2,000m. Flowers, 8-20mm across, are borne
on long stalks from December to February.

Family: Compositae

Mountain koromiko   *Hebe subalpina*
H. subalpina occurs in the southern mountains, commonly in the wetter
western areas. Flowers occur in great profusion from December to
February. (See also page 65.)

Family: Scrophulariaceae

258. Mountain koromiko in full flower, Arthur's Pass (December)

9. False buttercup, Homer Cirque, 1,000m (January)

0. Mountain pincushion in flower, Temple Basin, Arthur's Pass (January)

261. Comb-leaved pincushion in flower, Arthur's Pass (November)

Comb-leaved pincushion   *Cotula pectina*
A creeping, mat-forming herb with rather wiry, woolly stems beari
pinnatifid, comblike leaves 2-4cm long and 1-2cm wide. Flowers 6-8m
across are borne on long, woolly stalks from December to February. Fou
in the South Island mountains from Lewis Pass to Central Fiordlan
in damp rocky places, fellfields, herbfields and grasslands.

Family: Composit

Slender forstera   *Forstera tene*
A smooth-leaved herbaceous plant found in damp alpine grasslan
herbfields, or edges of bogs, from the Ruahine Ranges southwards
Stewart Island. The leaves, 6-10mm long, are distinctly separated on sh
petioles, and the flowers, 1cm across, on very slender stalks 5-10cm lor
occur from December to February. There are four species of the gen
Forstera in New Zealand. All bear delicate flowers on slender stalks
10cm long.

Family: Stylidiace

262. The slender forstera, Lewis Pass (January)

263. Slender forstera plants from Mt Holdsworth
    (February)

264. Leafy forstera in flower, Mt Egmont (January)

Leafy forstera   *Forstera bidwillii* var. *densifol*
Rather similar to *F. tenella* but with closely set branches, 4cm long, bearir
thick, densely imbricate leaves 6-10mm long by 4mm wide. Flowers,
15mm across, occur as one- to three-flowered clusters from Decembe
to March. Found in damp, alpine, rocky places, grassland and herbfiel
on Mts Egmont, Ruapehu and Tongariro between 800 and 2,000m.
                                        Family: Stylidiace

Soft herb   *Myriophyllum pedunculatu*
A soft aquatic or semi-aquatic herb, spreading by creeping and rootir
stems that produce simple erect branches, 2-10cm high, with thick, flesh
opposite leaves, 5-10mm long. Figure 265 shows a portion of a plant lifte
from amongst the gunnera. The fleshy leaves are clearly seen, with r
female flowers at the apex of the branch and male flowers on the lef
hand branch just below. Figure 266 shows the flowering branches of th
soft herb protruding from amongst solitary gunnera plants growing i
large patches beside Boulder Lake.
                                        Family: Haloragace

265. Soft herb, *Myriophyllum pedunculatum*, plant
in flower, Boulder Lake (February)

266. Soft herb, *Myriophyllum pedunculatum*, growing and flowering
amongst solitary gunnera, Boulder Lake (February)

267. A plant of Eyre Mountain daisy from Eyre Peak, growing in John Anderson's alpine garden at Albury (December)

Eyre Mountain daisy　　　　　　　　　　　　*Celmisia thomson*
A stout-stemmed, closely branched, spreading plant with its leaves arrange
as flat rosettes at the tips of the branches. The thick leaves are 10-15mr
long by 3-5mm wide, with fine bristles above and below; the flowers ar
up to 20mm across. Found in rocky crevices among the Eyre Mountain
between 1,650 and 2,000m.

Family: Composita

Red-fruited gunnera　　　　　　　　　　　　*Gunnera dentar*
A prostrate plant found along shady, subalpine streamsides and dam
places, forming mats up to 60cm across, with leaves 8-15mm long o
petioles up to 5cm long. Sometimes along river and streamsides sever
plants coalesce into extensive patches. On the banks of the Travers Rive
I have seen patches such as this extending over 2-3m; these, when th
drupes are ripe, make an attractive sight. Flowers occur during Novembe
and December, and the pendulous drupes, borne as open clusters o
elongated stalks, ripen during March and April.

Family: Haloragacea

Creeping gunnera　　　　　　　　　　　　*Gunnera proreper*
A creeping and rooting plant forming patches up to 60cm across. Th
leaves, 10-30mm long by 10-25mm wide, are on petioles 10cm long, an
the flowers appear from September to January on stalks up to 6cm high
The red, yellow, or green drupes, 3-4mm long, ripe by February, are th
only tight raspberry-shaped drupes occurring on any New Zealand gunner
Found from the Waikato south to Stewart Island, in damp herbfield
fellfields, grasslands and bogs, often associated with sphagnum moss, u
to elevations of 1,100m.

Family: Haloragacea

268. Red-fruited gunnera, Upper Travers Valley (April)

269. Creeping gunnera, Outerere Stream, base of Mt Tongariro (February)

270. Panakenake in flower, Mt Egmont (January)

271. Panakenake with berries, creeping over a low bank, Arthur's Pass (April)

anakenake   *Pratia angulata*
A slender, prostrate, creeping and rooting herb, often forming large mats
m across in damp shaded and sheltered places up to 1,500m. The leaves
ary in size and may be 3-12mm long and 1-8mm wide. The flowers,
-16mm across, occur in great profusion from October to March. The
erry, 8-12mm long, ripens from February to April.

Family: Lobeliaceae

72. Purple bladderwort, Key Summit (January)

urple bladderwort   *Utricularia monanthos*
A stemless plant, which sheds its leaves before flowering and which bears
iny bladders on its roots. These bladders, 1.5-2.5mm wide, have lids and
rap microscopic water animals that enter them. By digesting these animals,
he bladderwort, like a sundew, obtains nitrogen. Flowers appear during
anuary, apparently when the bog dries out. During the summer of 1967,
vhich was unusually dry in Fiordland, a marvellous flowering of this little
lant occurred throughout the alpine bogs of the whole area. The flowers
re about 1cm across on stems up to 10cm high.

Family: Lentibulariaceae

273. Green cushion daisy in flower on wet rocky face, 1,520m, Cupola Basin (January)

Green cushion daisy   *Celmisia bellidioide*

A creeping, much-branching and rooting daisy forming mats up to 1m across, found in the South Island mountains growing on wet rocks o gravel through which water is flowing. It occurs between 600 and 1,600m The fleshy, oblong to spathulate leaves are 8-12mm long by 3-4mm wide and the flowers, which appear during December and January, are 2cm across.

Family: Composita

274. Thistle-leaved senecio in flower, Mt Egmont (January)

275. Small-leaved wet rock hebe in flower, Boulder Lake (January)

Small-leaved wet rock hebe   *Parahebe lyallii*

A prostrate or decumbent, spreading and rooting shrub forming small carpets on wet rock and streamsides, up to 1,400m, among the mountains from the Ruahine Ranges to Fiordland. The rather thick but fleshy leaves are 5-10mm long. Flowers are borne as racemes on stalks up to 8cm long from November to March.

Although thriving in damp places, especially within the moist zone of a waterfall, when sprays of both the wet rock and streamside hebes are picked they drop the corollas of their flowers within a few moments if the stems are placed in water. The flowers, however, will remain intact on a spray for several days if it is kept in a plastic bag or even left in the open, dry. A spray kept in this way loses its corollas soon after being revived by placing the stems in water.

Family: Scrophulariaceae

Thistle-leaved senecio   *Senecio rufiglandulosus* var. *rufiglandulosus*

A leafy shrub reaching 1m high, with the lower leaves up to 10cm long by 5cm wide, deeply serrate or dentate. Flowers up to 2cm across occur in corymbs from November to February. Found along damp, shady streamsides and shady banks up to 1,250m altitude from the Volcanic Plateau southwards.

Family: Compositae

276. Wet rock hebe in flower, Arthur's Pass (December)

Wet rock hebe   *Parahebe linifolia*
A woody-based, much-branched shrub, sometimes rooting at the nodes, forming a sprawling mat up to 40cm across in wet rocky places and alongside streams up to 1,400m amongst the mountains of the South Island. The thick, sessile, narrow leaves are 8-20mm long by 1.5-4mm wide. The flowers, usually in two- to four-flowered racemes, appear during December and January.

Family: Scrophulariaceae

Streamside hebe   *Parahebe catarractae*
A low, variable, open-branched shrub with woody branches bearing leaves 1-4cm long by 5-20mm wide. Flowers 8-12mm across are borne profusely from November to April as few- or many-flowered racemes on long stalks. Found along damp streamsides, cliff faces and rocky places, in subalpine regions throughout New Zealand.

Family: Scrophulariaceae

Bush violet   *Viola filicaulis*
A glabrous, tufted herb with creeping, rooting stems rising at their tips to 15cm high. The smooth, crenulate leaves, 1-3cm long, are on petioles up to 2cm long with stipules up to 5mm long. Flowers 1-2cm across are borne on slender, flattened peduncles up to 10cm long, from October til January. Found throughout New Zealand in subalpine, damp, shaded places such as subalpine scrub and along seepages, edges of bogs, stream banks and open forest up to 1,500m.

Family: Violaceae

277. Streamside hebe in flower.
Tararua Ranges (December)

278. Bush violet flowers and leaves,
Wharite Peak (December)

279. New Zealand violet, Homer (December)

New Zealand violet   *Viola cunninghamii*
A glabrous, tufted herb up to 15cm high, found all over New Zealand
in damp places up to 1,600m altitude. The ovate, crenulate leaves, 1-3cm
long, are on hairy petioles up to 5cm long. The flowers, 1-2cm across,
arise on slender stalks 5-10cm high from October till January.

Family: Violaceae

Star herb   *Libertia pulchella*
A rather dainty little herb up to 12cm high, with grasslike leaves having
the ventral surface duller than the upper. Found throughout New Zealand
in damp, shady places in forests up to 1,100m. Bears flowers up to 15mm
across from November to January or February. The smooth, rounded
seeds that follow are yellow when fully mature.

Family: Iridaceae

Hairy ourisia   *Ourisia sessilifolia* var. *simpsonii*
A small *Ourisia* with hairy leaves, 1-2cm long, forming rosettes along a
creeping stem. The hairy flower stalk, 5-15cm high, bears flowers, 1-1.5cm
across, in pairs from December till February. The hairy ourisia is found
throughout the South Island mountains in damp grassland and wet fellfields
up to 2,000m altitude. The variety *simpsonii,* which has no hairs on the
lower leaf surface, or only on the veins, is found from the mountains
of northwest Nelson south to the Amuri Pass.

Family: Scrophulariaceae

280. Star herb in flower, Wharite Peak (December)

281. Hairy ourisia in damp grassland, Cupola Basin, 1,500m (December)

282. Broad-leaved elf's hood, Tararua Ranges
(November)

Broad-leaved elf's hood   *Pterostylis montana*
This orchid has stems up to 30cm high, broad leaves 1 cm wide, and
flowers 2-3cm long. It is distinguished from *P. banksii* by its broader
leaves and smaller flowers, which appear during November and December.
Found on shaded slopes in subalpine forest and scrub and in lowland
forest, from the Kaweka Ranges southwards to Stewart Island.

Family: Orchidaceae

Puatea   *Gnaphalium keriense*
A small, prostrate or decumbent, everlasting daisy, 15-24cm high with
spreading branches and sessile leaves 4-7cm long by 3-10mm wide. Flowers,
each about 1cm across, occur as flat corymbs on long, hairy stalks from
September till February. Found throughout the North Island and as far
south as North Canterbury in the South Island, growing in subalpine regions
along shady streamsides and banks, and in lowland areas sometimes
clothing shaded roadside banks.

Family: Compositae

283. Tutukiwi flowers, Tararua Ranges (November)

utukiwi *Pterostylis banksii*

n orchid with narrow leaves 6-12mm wide and with stems 30-45cm high,
hich bear flowers 5-7.5cm long including their tails. The flowers occur
om September to December, and the plant is found all over New Zealand
damp and shady places up to 1,250m altitude.

Family: Orchidaceae

284. Puatea in flower, Outerere Stream, base of Mt Tongariro (December)

Lantern berry   *Luzuriaga parviflo.*
A delicate, creeping lily found throughout New Zealand on damp, mos
covered banks and moss-covered tree trunks in subalpine forests and scru
up to 1,100m altitude. The alternate, sessile leaves, 1.5-4cm long, are thic
and rigid. The white flowers, up to 2cm across, occur from Novembe
to February, and the berries mature from January to March.

Family: Liliace;

285. Lantern berry with flowers and berry, Mt Egmont, 1,000m (December)

Tufted grass cushion   *Colobanthus apetal*
A loosely tufted cushion plant with thick, grass-like, flexuous, pointe
leaves, 10-25mm long. Flowers up to 6mm across, composed of five sepa
occur during December and January. The seeds mature during Marc
and April, when the capsules open displaying these small, round seed
Found from Mt Hikurangi southwards in wet, sandy, peaty or mudc
places up to 1,250m altitude.

Family: Caryophyllace;

6. Tufted grass cushion with fruiting capsules, Kaimanawa Ranges (November)

7. Close-up view of seed capsule of tufted grass cushion (November)

288. Alpine grass cushion, growing in rocky crevice, Fog Peak, 1,650m
(January)

289. Close-up view of seed capsule of alpine grass
cushion (November)

reeping ourisia   *Ourisia caespitosa* var. *gracilis*
prostrate, much-branched, creeping plant, rooting at the nodes, found
shady, damp places amongst the mountains of Canterbury and Otago
p to 1,300m. The leaves, 6mm long by 3mm broad, are spathulate with
nooth margins. The flowers, 1.5cm across, occur from October till
ecember, mostly on hairless stalks 8-10cm long.

Family: Scrophulariaceae

90. Creeping ourisia var. *gracilis*, Mt Belle, Homer (December)

lpine grass cushions   *Colobanthus* spp.
)ense-tufted cushion plants bearing stiff, erect, pointed leaves up to 25mm
)ng. Flowers up to 15mm long occur from October to April, and the
:ed capsules mature and open one year later. Found up to 1,700m, mainly
wet rocky crevices, wet shingly places or damp grasslands from the
.uahine Ranges southwards but only on the east side of the South Island
ountains. Of the 13 grass cushion plants found in New Zealand, all
ut one occur in the mountainous regions and figures 288-9 show *C.
uchananii*, a typical example of these.

Family: Caryophyllaceae

291. Creeping ourisia, Travers Valley (December)

Creeping ourisia   *Ourisia caespitosa* var. *caespitos*
Plant similar to variety *gracilis* but with the leaves 10mm long by 5mm
wide, broadly spathulate with notched margins. Flowers 2cm across occu
from November to February, mostly in pairs on hairless stalks up to 10c
long. Found throughout New Zealand in shady, damp and wet, rock
subalpine and alpine places up to 1,500m altitude.

Family: Scrophulariacea

North Island mountain foxglove   *Ourisia macrophyl*
Plants arise from a prostrate rooting stem as terminal tufts of leaves 4
10cm long. The stout, erect flower stalk, up to 60cm high, usually carrie
at least one whorl of small cauline leaves below the first flower cluster
Several varieties of the plant are recognised, and the variety *macrophyl*
(fig. 293), found only on Mt Egmont above the bushline, has its leave
not much longer than their width. The variety *robusta* (fig. 292), foun
on the Volcanic Plateau and between Taupo and Napier, has leaves o
distinctly greater length than width. Mountain foxgloves grow in shady
damp, subalpine situations and flower from October till January o
February. On Mt Egmont many of the banks along the walking track
have become clothed with this *Ourisia,* and during December and Januar
when the plants are in full flower these tracks are a real delight to see.

Family: Scrophulariacea

292. Plants of North Island mountain foxglove var.
robusta, near Mt Tongariro (December)

293. Flowers of North Island mountain foxglove var. *macrophylla*, Mt
Egmont (January)

294. Plant of South Island mountain foxglove, var. *calycina*, Mt Temple, Arthur's Pass, 1,500m (January)

295. Flowering plants of South Island mountain foxglove, var. *calycina*, growing on tussock-shaded bank, Mt Temple, Arthur's Pass, 1,500m (January)

Hairy ourisia *Ourisia macrophylla* var. *lactea*

This is a variety of *O. macrophylla*, the North Island mountain foxglove, found in damp herbfields or grasslands of the South Island up to 1,850m, from Lewis Pass southwards. It is a shorter plant than *O. macrophylla* (fig. 293), having leaves 2cm long and 2cm wide, and it flowers during December and January.

Family: Scrophulariaceae

296. Hairy ourisia in flower in tussock herbfield, Mt Lucretia, Lewis Pass, 1,300m (January)

South Island mountain foxglove *Ourisia macrocarpa* var. *calycina*

The plants form tufts of leaves from creeping stems as in *O. macrophylla*. The leaves, 4-15cm long by 2-10cm wide, are on long, hairy petioles, usually larger than the leaf. Stout, hairy flower stalks up to 70cm high bear whorls of flowers from October to January. Found throughout the South Island mountains in damp places in scrub, herbfields and fellfields. Two varieties are recognised, variety *calycina* occurring in the mountains north of Franz Josef and variety *macrocarpa* in the southern mountains.

Family: Scrophulariaceae

297. Mountain astelia in grassland, Cupola Basin, 1,600m
(December)

Mountain astelia   *Astelia nervosa*
A tussock-like plant that ranges from small to quite large, with silvery,
hairy leaves up to 75cm long. The ribs of the leaves are often coloured
red or purple. Found all over the New Zealand mountains in damp places
in grasslands, herbfields and fellfields between 750 and 1,400m. The flowers
(fig. 299), which are strongly sweet-scented, appear during November,
December and January. The berries, 8mm in diameter (fig. 298), ripen
from March to May. This lily often occurs in large colonies covering quite
extensive areas on mountainsides. Specimens of mountain astelias with
bronze-coloured leaves probably belong to a different species.

Family: Liliaceae

298. Berries of mountain astelia, Mt Holdsworth (May)

299. Flowers of mountain astelia, Mt Egmont, 1,300m (December)

300. Kakaha showing wide nerves of the leaves and partially open fruiting heads, Mt Hector 1,050m (March)

301. Ripe berries of Kakaha, Mt Hector (March)

Small feathery-leaved buttercup   *Ranunculus gracilipes*

A delicate herb up to 15cm high, with finely pinnate or divided leaves
on hairy petioles 2-12cm long. Flower stalks are sometimes hairy, up to
5cm high, each with a single flower about 2cm across with 5-8 petals.
Flowers occur during November and December and this buttercup is found
in the South Island in damp alpine grasslands and herbfields alongside
streams from Mt Travers southwards.

Family: Ranunculaceae

302. Small feathery-leaved buttercup by a stream in grassland, Cupola Basin, 1,500m
(December)

Kakaha   *Astelia fragrans*

This lily is easily recognised by its large leaves, 60cm-2.6m long and 2-
10cm wide, with their clear nerves running down each side of the midribs.
Sweet-scented flowers occur on branching stalks up to 60cm long from
October to January. The fruit takes almost a whole year to mature, forming
orange berries sitting in yellow cups, crowded together on a stout, branching
stalk 25-30cm long. The yellow cups seen in figure 300 are also a
distinguishing feature of this astelia, which is found in damp situations
in forest or scrub up to 1,100m throughout New Zealand.

Family: Liliaceae

303. Dense sedge with flower spikes, Tauhara Mountain, 850m
(December)

304. Pigmy bamboo grass, Cupola Basin, 1,500m
(April)

Dense sedge   *Uncinia uncinata*
A densely tufted sedge, with leaves 6-8mm wide forming clumps up to
30cm high in shaded, open places up to 900m altitude. It is common
in clearings in subalpine forest or scrub and along bush tracks. Flowers
appear from November to February as dense spikes 15cm high. The black
seeds, ripe from March onwards, cling to the clothing and skin of passersby.
                                          Family: Cyperaceae.

Pigmy bamboo grass   *Microlaena colensoi*
A tufted grass, branched at the base and reaching 15-45cm high, with
leaves 10-15cm long by 4-6mm wide. Found in damp situations between
900 and 1,700m from the Ruahine Ranges southwards to Fiordland.
                                            Family: Gramineae

305. Broad-leaved sedge in flower, Taupo
(December)

Broad-leaved sedge   *Machaerina sinclairii*
Easily recognised by its broad leaves, 1.5-2.5cm across and 1-1.3m long,
and large drooping flower heads, this sedge is found on shaded banks
and streamsides in the North Island from Wanganui northwards and on
the Ruahine and Kaimanawa Mountains up to 800m.
                                          Family: Cyperaceae

HERBFIELDS AND FELLFIELDS

HERBFIELDS

Herbfields occur mainly on the higher slopes above the tussock grassland and consist of large herbaceous plants growing in abundance with, in some places, an intermingling of low shrubs. Fellfields is the name given to more open plant associations found in unstable rocky places in either wet or dry situations. Fellfield plants are mostly low-growing forms capable of thriving on an unstable substrate in a harsh climate.

Large feathery-leaved buttercup *Ranunculus sericophyllus*
A perennial herb with numerous, deeply divided, hairy leaves on petioles
2-12cm long, found mainly in higher alpine, wet, rocky places and fellfields
between 1,550 and 2,150m, from the Lewis Pass to Northern Fiordland.
Flowers, 2.5-4cm across with five petals, occur from December to February.
Family: Ranunculaceae

307. Large feathery-leaved buttercup, Homer Saddle, 1,600m (January)

Giant buttercup *Ranunculus lyallii*
A perennial herb, often referred to as the 'Mount Cook Lily', which grows
to a height of 1.5m with leaves 13-20cm across. Flowers are produced
in profusion as large panicles during December and January. This is the
largest ranunculus plant in the world and is found in subalpine and alpine
herbfields and creeksides from Marlborough to Fiordland. Although
comparatively common, it is local in occurrence. It makes an unforgettable
sight when found in quantity clothing entire slopes with its sparkling flowers.
Family: Ranunculaceae

306. Giant buttercup, *Ranunculus lyallii*, Arthur's Pass (January)

308. Korikori plant in flower, Mt Ruapehu, 1,550m (January)

309. Flowers of lobe-leaf buttercup on long scapes showing cauline leaves, Mt Holdsworth, (December)

310. Korikori flowers, Mt Holdsworth, 1,250m (December)

Korikori   *Ranunculus insignis*

A branching, hairy herb up to 90cm high, found in alpine grasslands, herbfields and rock crevices from East Cape to the Kaikoura Mountains. The large thick leaves, 10-16cm wide, are clothed with silky hairs on their lower surfaces. Flowers 2-5cm across occur in profusion from November to February, and plants in full flower among the rocky funnels of Ruapehu or Holdsworth are a glorious sight. A somewhat similar plant having distinctly reticulate, veined leaves without hairs, *R. godleyanus* occurs in the South Island mountains in similar situations from the vicinity of the Lewis Pass to Mt Cook.

Family: Ranunculaceae

Lobe-leaf buttercup   *Ranunculus verticillatus*

An erect but lax and slender buttercup with only a few leaves, 5-10cm across, usually deeply lobed, often divided almost to the base, and on slender petioles 7-15cm long. Flower stalks may be short or up to 60cm long (fig. 309). Long stalks are often supported by the vegetation surrounding the plant. The flowers, which occur sparingly, two or three per plant, are 2-3cm across with numerous petals having rounded tips, and appear during December and January. Found in wet situations among rocks, scrub and fellfields between 800 and 1,550m from the Ruahine Mountains south through Mt Hauhangatahi and the Tararuas to the mountains of northwest Nelson. Quite common on Mt Holdsworth, where it flowers well during December.

Family: Ranunculaceae

311. Snow buttercup, Mt Egmont (December)

Snow buttercup *Ranunculus nivicolus*

An erect, more or less hairy, perennial herb producing flowering branches up to 80cm high. The leaves, 5-15cm wide, on hairy petioles 10-20cm long, are usually deeply lobed with reticulate veins. The flowers, 3-5cm across, occur from December to February, and the plant is found on Mt Egmont, the Central volcanoes, and the Kaweka and Raukumara Ranges between 1,200 and 1,850m in subalpine scrub, fellfields and on finely divided scoria slopes. On Mt Egmont it reaches to the limit of the perpetual snowfields.

Family: Ranunculaceae

Grassland buttercup *Ranunculus* sp. (unnamed)

A tufted buttercup, similar to *R. multiscapeus*, up to 15cm high, with rosettes of leaves 1-2cm wide on hairy petioles up to 5cm long. From October to March, flowers, each of five petals and 1-2.5cm across, are borne on stalks that are usually hairy but may be hairless. The typical plant is a single rosette bearing one flower, but plants with many flowers arising from many branching root stocks are frequently met with. Found all over New Zealand in grassland and herbfields up to 1,750m. During November and December, when flowering is at its peak, this little buttercup often carpets riverflats with myriads of golden-yellow flowers imparting a brilliant glow to the landscape on a sunny day. The flower stalk, which in subalpine and lower regions is tall, up to 15cm, becomes progressively shorter as higher altitudes are encountered; at about 1,700m it virtually disappears, and the stalkless flower arises from the centre of the plant. I have seen this transition from the Travers Valley to Cupola Basin where growing amongst the tussocks from about 1,700m upwards, each plant has a stalkless flower.

Family: Ranunculaceae

312. Typical form of grassland buttercup, slopes of Sugarloaf, Cass (December)

313. Grassland buttercup with several flowers, Tasman Glacier Moraine (December)

314. Flowers of small variable buttercup, Porter's
Pass (November)

315. Clump of small variable buttercup with deeply divided leaves,
Porter's Pass (November)

316. Tall pinatoro flowers, Mt Ruapehu (December)

Tall pinatoro   *Pimelia buxifolia*

A rather rigid, much-branching shrub up to 1m high, found in fellfields
and grasslands amongst the North Island mountains. The leaves, 5-10mm
long by 3-5mm wide, are keeled. Flowers occur from about September
on until February.

A very similar plant, *P. traversii*, found in fellfields and rocky places
throughout the South Island, is up to 60cm high, with sessile, faintly keeled
leaves 4-9mm long by 3-6mm wide, black bark, and very hairy, pinkish-
coloured flowers.

Family: Thymelaeaceae

Small variable buttercup   *Ranunculus enysii*

A small, rosette-forming, hairless buttercup of very variable leaf form and
size. Sometimes appressed to the ground, sometimes up to 15cm high,
with flowering stems 10-40cm high. The leaves, 2-10cm long by 1-7cm
wide on slender, grooved petioles up to 10cm long, are variously divided
and lobed. Flowers 1.5-3cm across, with normally fine petals, appear during
October and November. Found in sheltered places among tussocks and
scrub or in rocky clefts along the eastern mountains of the South Island
from North Canterbury to southwest Otago.

Family: Ranunculaceae

317. Plant of North Island edelweiss in flower, Mt Holdsworth, 1,450m (February)

North Island edelweiss *Leucogenes leontopodium*
A small woody plant with branching, decumbent stems curving upwards
at their tips and bearing sessile leaves 8-20mm long by 4-5mm wide, which
are covered on both surfaces with dense, silvery-white wool. Flower heads,
5.5-6.5cm across, are terminal, one to each ascending branch, and they
occur from November until March. The flower is formed of 8-15 central
clusters surrounded by a ring of up to 20 white, densely woolly leaves.
Found in rocky alpine herbfields and fellfields, often in crevices in rocks,
above 1,250m from East Cape to mid-Canterbury.

Family: Compositae

NEW ZEALAND GENTIANS

Nineteen species of gentians are known from the mountains of New Zealand, excluding the outlying islands. They flower mostly in late summer or autumn and all belong to the family Gentianaceae.

South Island mountain gentian   *Gentiana montana*

A thick-leaved gentian, reaching 50cm high, usually with several stout stems arising from a woody stock. Leaves are normally 1.5-2cm long by 10-15mm wide, but sometimes larger. White flowers, as terminal cymes of about 10 flowers each, occur during February and March. Each corolla is cut over halfway down into distinct rounded or angled lobes. Found in grasslands and fellfields or in the open throughout the South Island mountains.

218. Plant of South Island mountain gentian growing through a *Celmisia*, Mt Lucretia, 1,500m (January)

319. Flower stem of alpine gentian, Mt Hector
     Track, 1,500m (March)

320. Alpine gentian flowers, Mt Holdsworth, 1,400m (February)

Alpine gentian   *Gentiana patula*

A perennial herb with sprawling stems rising at their tips to a height of
10-50cm to bear cymes of lovely flowers, each 20-25mm across, during
January and February. The basal leaves are arranged as a rosette, oblong,
lanceolate, or spathulate in shape, and 2-7.5cm long by 1-2cm wide. Found
from the Tararua Ranges southwards in alpine and subalpine herbfields
and grasslands. During February it is not uncommon to find large areas
of this gentian in bloom on slopes in the vicinity of Mt Hector.

Small New Zealand gentian   *Gentiana grisebachii*

A slender annual gentian with weak decumbent to ascending stems, 7-
20cm long, and thin spathulate or oblong-spathulate leaves 15-20mm long
by 8-10mm wide. Flowers that are solitary or paired and about 12mm
long occur during January and February. Found in damp places in
herbfields, subalpine scrub and grasslands up to 1,300m altitude from Mt
Ruapehu southwards.

321. Flower of small New Zealand gentian,
Volcanic Plateau (February)

322. Flowers of Northwest Nelson gentian, Cupola Basin, 1,450m (April)

323. Northwest Nelson gentian, Brown Cow Pass (February)

Nelson alpine gentian *Gentiana vernicosa*

A perennial plant giving rise to one to five stems, 10-25cm long, each bearing two- to seven-flowered terminal umbels during February and March, with each flower 12-20mm across. The narrow basal leaves, up to 35mm long and 8mm wide, are in compact tufts. Found in herbfields and grasslands amongst the northwest and southwest Nelson mountains.

324. Nelson alpine gentian in flower above Boulder Lake, 1,550m (February)

Northwest Nelson gentian *Gentiana spenceri*

An annual gentian with a single or several stems, usually up to 15cm tall. Basal leaves 2-4cm long. Flowers, about 2cm across, occur from February to April as umbels surrounded by a whorl of five to seven oblong-spathulate leaves. Found throughout the mountains of northwest Nelson and northern Westland, in subalpine scrub, herbfields and grasslands, preferably in shaded situations that remain moist.

Pink gentian   *Gentiana tenuifoli*
A perennial plant with single, erect, finely-ridged stems up to 40cm high
The basal leaves are thin, obovate to spathulate, 8-14cm long by 2-3.5cr
wide, usually acute and crowded in rosettes. Flowers up to 2.5cm acros
occur during February and March, and the plant is found from northwes
Nelson to the Lewis Pass in subalpine herbfields and grasslands.

Ridge-stemmed gentian   *Gentiana tereticaul*
An erect annual herb with slender, ridged and lined stems 25-45cm high
The few thin, reticulated, ovate basal leaves are 25-35mm long and 1(
16mm wide on flat petioles 15-20mm long. Flowers, up to 2cm acros
are both terminal and subterminal in umbels. Upper cauline leaves ar
sessile and cuneate, usually in pairs. Primarily a Fiordland gentian, i
exact distribution is uncertain.

Common New Zealand gentian   *Gentiana bellidifoli*
A perennial herb with branching rootstock giving rise to crowded tuf
of thick, somewhat fleshy, more or less elliptic or spathulate, basal leaves
10-15mm long by 5-7mm wide. There are usually several either simpl
or branched flower stems, 5-15cm high, bearing two- to six-flowere
terminal cymes with each flower about 2cm across. Found from M
Hikurangi southwards to Fiordland in subalpine boggy places, or dam
herbfields and grasslands.

325. Flowers of common New Zealand gentian, Arthur's Pass (April)

326. Pink gentian flowers, Upper
     Cobb Valley (February)

27. Ridge-stemmed gentian in
    flower, Cupola Basin (April)

Townson's gentian   *Gentiana townson*
A perennial herb with a single or several, occasionally branched, slender
finely-lined stems up to 30cm high. Basal leaves fleshy and crowded, o
varying shape, 20-30mm long and 8-10mm wide and more or less glossy
Lower cauline leaves are similar, but the upper ones are small, thick an
tapering. Flowers occur during January and February, and the plant i
found in subalpine, damp places, amongst herbfields and grassland fror
Northwest Nelson to Arthur's Pass.

328. Flowers of Townson's gentian, Boulder Lake
(January)

Snow gentian   *Gentiana matthews*
A tall, slender gentian, branched and dividing from the base. The bas
leaves are 1-3cm long by about 6mm wide, thin, somewhat spathulat
and narrowing to a flat petiole 1-2cm long. The flowers, normally whit
but sometimes tinged with mauve, are produced in profusion from Januar
to March, are up to 2.5cm across and arranged, usually in twos or threes
at the tips of the branches.

329. Large plant of snow gentian coming into flower, Gertrude
Cirque, Homer, 900m (January)

330. Mauve form of snow gentian, McKenzie Pass
(March)

331. Plant of Benmore gentian, slopes of Mt Benmore, Upper Ure River, 600m (early March)

332. Flowers of Benmore gentian, Upper Ure River Valley, 700m (late March)

333. Normal plant of tall gentian, McKenzie Pass
     (March)

Tall gentian   *Gentiana corymbifera*

The willow-like basal leaves, up to 15cm long and 20mm wide, form rosettes,
and the tall flower stalks rise up to 60cm, bearing flowers 12-20mm across.
The stem leaves, also willow-like, are up to 7.5cm long. Found throughout
the mountains of the South Island, in herbfields and grasslands up to
1,250m. Flowers occur from January to March and are normally white,
but mauve-tinted varieties can occur.

Benmore gentian   *Gentiana astonii*

A spreading, erect gentian forming large plants up to 1m across; found
growing only on limestone outcrops in montane herbfields and grasslands
along the river valleys of the Northern Kaikoura Coast from the Clarence
to the Ure. Stems up to 45cm long; leaves small, in pairs, 2cm long and
1-2mm wide. Flowers occur in profusion during March and April, each
flower on a slender pedicel about 10mm long.

334. Mountain pratia in flower, Arthur's Pass (January)

335. Mountain pratia with berries, creeping over a low bank, Arthur's Pass (April)

Mountain pratia *Pratia macrodon*

A prostrate, creeping and rooting herb, found in subalpine and alpine herbfields, fellfields, exposed rocky places and grasslands throughout the South Island mountains between 900 and 1,600m. The strongly toothed leaves are 5-8mm long. The flowers, 12-18mm long, occur from November to March and are followed during March and April by ovoid, purple-coloured berries, 6-9mm long. Distinguished from *P. angulata* by more deeply toothed and thicker leaves and the tapering, pointed petals of the flowers.

Family: Lobeliaceae

Trailing celmisia *Celmisia ramulosa*

A trailing plant with stiff, erect branches densely clothed with thick, overlapping leaves, 5-10mm long by 1-2mm wide. The leaf margins are strongly recurved and the lower surface clothed with a dense, white tomentum. Flowers, 2-2.5cm across, occur during November and December on stout woolly stalks arising from the branch tips and each bearing a few small, narrow bracts. Found mainly trailing over rocky places or in herbfields up to 1,400m in the mountains of central and western Otago, Southland and Fiordland.

Family: Compositae

336. Trailing celmisia in flower, above Wapiti Lake, Fiordland (December)

337. Plant of scarlet snowberry in flower, slopes of Mt Belle, Homer (December)

Scarlet snowberry   *Gaultheria crass*
A much-branched, rather untidy shrub up to 1m high, which become
dwarfed at higher altitudes. The thick, leathery leaves, 10-15mm long b
5-7mm wide, are on stout petioles 1mm long. Flowers, distinguished b
their reddish-tipped sepals, occur as racemes up to 4cm long from Octobe
to December, and the berries, 2-3mm across, colour during February
March and April. Found from the Ruahine Mountains southwards i
subalpine herbfields, fellfields and grassy and rocky places.

Family: Ericacea

338. Flowers of scarlet snowberry, Homer (December)

339. Berries of scarlet snowberry, Jack's Pass (March)

340. White berries of mountain snowberry, *G. depressa* var. *novae-zelandiae*, Cupola Basin (January)

341. Pink berries of mountain snowberry, *G. depressa* var. *novae-zelandiae*, Boulder Lake, 1,250m (January). *Gaultheria* species hybridise with one another and with species of *Pernettya* (figs. 118-120.)

Alpine avens   *Geum uniflorum*

One of the four endemic avens, this herbaceous plant often forms large patches in damp areas of subalpine and alpine herbfields and open rocky places from the Nelson mountains to Otago, especially on the wet side of the Alps. The rounded, hairy-margined leaves, up to 10cm long, arise directly from the rootstock. Flowers up to 2.5cm across occur singly during January and February.

Family: Rosaceae

242. Alpine avens in flower, Temple Basin, Arthur's Pass (January)

Mountain snowberry   *Gaultheria depressa* var. *novae-zelandiae*

A prostrate or decumbent, creeping and rooting shrub with many interlacing branches and hairy branchlets bearing thick, leathery, crenulate leaves, 4-10mm long by 4-6mm wide, on very short petioles. The plant, seldom more than 10cm high, is common in subalpine and alpine herbfields, fellfields, rocky and boggy places, and grasslands from the Kaimanawa Mountains south to Stewart Island. Flowering starts during November in the north and continues till February in the south. The berries, 3-4mm across, mature from January onwards and may be white or pink.

Family: Ericaceae

343. Common cardamine plant in full flower, Temple Basin (January)

Cardamines *Cardamine* spp

Soft herbs forming mats or tufts up to 30cm across in forests, along the banks of streams and near stones in herbfields and fellfields throughout New Zealand. Flowers about 5-6mm across occur in profusion during December and January.

Family: Cruciferae

Common lycopodium *Lycopodium scariosum*

Mountain lycopodium *Lycopodium fastigiatum*

Found all over New Zealand and extending into subalpine scrub and herbfields, these plants have creeping stems up to 2m long with ascending branches up to 30cm high. The flattened leaves of *L. scariosum*, about 3mm long, tend to lie opposite in pairs on one plane. The strobili or fruiting bodies, 2.5-5cm long, occur at the tips of the branches, usually singly, from December to April. The closely imbricate leaves of *L. fastigiatum* are 3-4mm long, and the terminal strobili, 2-5cm long, occur singly, in pairs or several together from November to March.

Family: Lycopodiaceae

344. Fruiting bodies of common lycopodium, Arthur's Pass (April)

345. Mountain lycopodium with fruiting bodies, Mt Holdsworth (February)

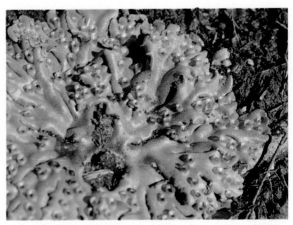

346. Giant liverwort with gemmae cups, Cupola Basin, 1,450m (April)

Mountain mikimiki   *Cyathodes empetrifolia*
A prostrate shrub with slender, wiry branches up to 40cm long and very
slender, hairy branchlets up to 15cm long, bearing thick, narrow leaves
3-5mm long, with recurved margins. The small, axillary or terminal flowers
occur from November to February, and the drupes, 4-5mm in diameter,
mature from January till April. Found in subalpine and alpine herbfields,
fellfields, grasslands and boggy places throughout New Zealand.

Family: Epacridaceae

347. Mountain mikimiki with berries, Volcanic Plateau (February)

Giant liverwort *Marchantia* spp.

Liverworts are a simple form of plant found on rocks in the spray of waterfalls, and in other damp places such as wet rocks, moist soil, rotten logs and damp tree trucks in lowland and alpine regions. This fine specimen of a liverwort, some 7cm across, is covered with gemmae cups in which are formed reproductive buds that will give rise to new plants. Liverworts are the most simple of land plants and they can live only in continuously moist situations. They have neither leaves nor roots and grow solely by peripheral enlargement of the thallous, or plant body, shown here.

Family: Marchantiaceae

Mountain heath *Leucopogon suaveolens*

A prostrate or decumbent, branching shrub forming extensive patches or dense hummocks from 8cm-1m high and up to 2m across. The leaves and branches are normally reddish coloured, but sometimes bluish-green plants may be encountered (fig. 351). The leaves, 5-9mm long, are characterised by five, conspicuous, parallel veins on each undersurface and the margins are hairy. Flowers occur as two-to five-flowered terminal racemes, from November to February, each flower being about 7mm long. The fruit, a small round drupe about 3mm in diameter, which may be white, pink or crimson, matures from January to April. Found from the Volcanic Plateau southwards in herbfields, fellfields, grasslands and exposed rocky places.

Family: Epacridaceae.

348. Berries of mountain heath, Volcanic Plateau (February)

349. Close view of flower buds, flowers and leaves of mountain heath, Sugarloaf, Cass (November)

350. Mountain heath with red foliage, showing new shoots and flowers, spreading over roc Sugarloaf, Cass, 1,250m (November)

351. Spray of mountain heath in flower with green foliage,
Moor Park, Abel Tasman National Park (November)

352. Patotara in flower, Cass, 1,000m (November)

Patotara   *Leucopogon fraser*

A spreading, shrubby plant with prostrate, decumbent or ascending
branches, up to 15cm high, forming extensive, more or less dense patches
in subalpine fellfields, exposed rocky places and grasslands. The leaves,
4-9mm long, have pungent tips about 2mm long. Flowers, about 1cm
long, occur in profusion from September to January, and the drupes mature
from February till April.

Family: Epacridaceae

353. Patotara berries, Cupola Basin, 1,550m (April)

354. New Zealand eyebright in flower, Mt Holdsworth, 1,250m (February)

New Zealand eyebright   *Euphrasia cuneata*

A perennial herb up to 60cm high. The smooth, wedge-shaped leaves, 5-15mm long, have slightly thickened, flat margins and petioles 1-4mm long. The flowers, 15-20mm long, occur in abundance from January to March. Found in herbfields, fellfields, subalpine scrub, damp, rocky places and along streamsides up to 1,550m altitude, throughout New Zealand to as far south as North Canterbury.

Family: Scrophulariaceae

355. Flowers of New Zealand eyebright, Mt Hector (March)

356. Tararua eyebright in flower, Mt Holdsworth, 1,450m (December)

357. Flowers of Tararua eyebright, Mt Holdsworth (December)

358. Lesser New Zealand eyebright in flower, Homer Cirque (January)

Lesser New Zealand eyebright   *Euphrasia zelandica*
A sparingly branched, somewhat succulent, hairy, annual herb, 5-20cm high. The hairy, sessile leaves, 4-10mm long by 2-6mm wide, have thickened, recurved margins and tend to cluster as rosettes at the tips of the branches. Flowers about 10mm long occur during January and February. Found in damp places in herbfields and fellfields from Mt Hikurangi southwards. A very similar but smaller eyebright, usually without hairs on the leaves, *E. australis*, is found throughout the Fiordland mountains.

Family: Scrophulariaceae

Tararua eyebright   *Euphrasia drucei*
A tufted, creeping and rooting, perennial herb with erect branches up to 5cm high. The close-set leaves, 2-10mm long by 1-5mm wide, are sessile with thickened margins rolled slightly outwards. Flowers 10-15mm across occur in abundance from December to February, and the plant is found in boggy places in alpine herbfields, fellfields and grasslands from the Ruahine Mountains southwards to Fiordland.

Family: Scrophulariaceae

359. Yellow eyebright in flower, Arthur's Pass (January)

Yellow eyebright   *Euphrasia cockayniana*
A small, succulent herb, 5-10cm high, somewhat similar to *E. australis*
in general appearance but producing bright-yellow flowers 12-14mm long,
either singly or in pairs, towards the tips of the branches from December
to March. Found in the South Island in alpine herbfields and boggy places
of the Paparoa Mountains and the vicinity of Arthur's Pass.
                                        Family: Scrophulariaceae

360. Clustered eyebright in flower, Cupola Basin, 1,600m (December)

361. Alpine heath in flower, summit Tauhara
Mountain (December)

Alpine heath   *Epacris alpina*

A compact, erect, heath-like shrub, up to 1m high, with erect or spreading
branches. The thick, leathery, pointed, close-set leaves, 3-5mm long and
about 2mm wide, are concave-convex and keeled on the lower surface.
Flowers about 7mm long occur in profusion along the terminal portions
of the branches during December and January. Found in subalpine
herbfields, fellfields, scrub and grassland from the Volcanic Plateau to
Lake Tekapo.

Family: Epacridaceae

Clustered eyebright   *Euphrasia monroi*

A perennial herb with the lower stems usually leafless and the upper stems
densely leafy. The leaves are sessile, close-set and spreading, 5-10mm long
by 3-6mm wide, with thickened margins. Flowers 10-15mm long occur
as clusters at the tops of the stems from October to February. Found
among the mountains of Nelson and Marlborough in herbfields, usually
between 1,250 and 1,600m. A very similar plant, *E. townsonii*, occurs
among the mountains of northwest Nelson. It is distinguished by having
its leaves more separated and its flowers occurring singly on slender stalks
5-20mm long.

Family: Scrophulariaceae

Everlasting daisy   *Helichrysum bellidioides*

A prostrate, creeping and rooting shrub with stems up to 60cm long. Branches are woolly and the leaves, 5-6mm long by 3-4mm wide, are on short petioles and are softly woolly below. Flower heads 12-15mm across are produced in great profusion on long, woolly stalks from October to February. Found from the Coromandel Mountains south to Stewart Island in subalpine herbfields, fellfields, exposed rocky places and grasslands.

Family: Compositae

362. Plant of everlasting daisy, Arthur's Pass (January)

Common drapetes   *Drapetes dieffenbachii*

A prostrate, sprawling plant, forming dense or open patches up to 30cm across in alpine fellfields and grasslands from the Coromandel Mountains south to Stewart Island. The leaves are generally appressed to the branchlets, are 2.5-3.5mm long, and taper from base to apex. Flowers, two to four per cluster, occur terminally on the branchlets from November to January.

Family: Thymelaeaceae

363. Common drapetes in flower pushes out from under a *Celmisia* and tussock, Mt Holdsworth (December)

364. Close view of common drapetes stem, Mt Ruapehu (December)

365. Spray of Mason's swamp hebe, Lewis Pass (December)

Swamp hebe   *Hebe pauciramosa*
The principal variety of the swamp hebe is a little-branched, erect shrub,
up to 50cm high, found in damp places in herbfields and grasslands among
the mountains from Canterbury to Southland. The keel of the leaf flattens
out below the leaf tip, but in the variety *masonae* this keel is complete
to the leaf tip.

Family: Scrophulariaceae

MOUNTAIN DAISIES

Mountain daisies belonging to the genus *Celmisia*, family Compositae, are among the most common plants in the New Zealand mountains. A selection of these is shown in the following pages.

66. Armstrong's daisy in flower, Arthur's Pass (January)

Armstrong's daisy   *Celmisia armstrongii*
A large, tufted daisy in which the rigid, thick, leathery, longitudinally ridged leaves, 20-35cm long by 1-2cm wide, have a characteristic yellow- or orange-coloured band running down each side of the midrib. Flowers 4-5cm across on stalks 25cm long occur during January and February. Found in alpine herbfields and grasslands amongst the mountains of the South Island.

C. lanceolata occurs in similar situations but differs in that the midrib is stout and yellow to orange in colour.

367. Large mountain daisy showing the beautiful symmetry of the flower and its bud, Arthur
Pass (December)

368. Large mountain daisy, Arthur's Pass (January)

Large mountain daisy, tikumu, silvery cotton plant

Celmisia semicordata

A large, robust, tufted herb with leaves more or less silvery above when young, dark green when older, and reaching a length of 60cm. Flower heads 5-12cm across occur on fluffy, woolly stems up to 75cm high from December to February. Flowers of plants from the Arthur's Pass region tend to have drooping petals (fig. 368), while those of plants from further south have upright petals and more silvery leaves (fig. 369). This, the largest of the mountain daisies, is found in alpine herbfields, fellfields and grasslands throughout the South Island.

369. Large mountain daisy with very silvery leaves, Homer Cirque (December)

370. Plant of musk daisy showing leaves, Mt Lucretia, Lewis Pass, 1,300m (January)

371. Plant of musk daisy in full flower, Arthur's Pass (January)

372. Strap-leaved daisy in flower, Fog Peak, Porter's Pass, 1,680m (January)

Strap-leaved daisy *Celmisia angustifolia*

A small, woody shrub, with lower branches clothed by dead leaves and bearing tufts of rosulate, living leaves at their tips. The leathery leaves are straplike, 3-5cm long by 3-5mm wide, sticky with soft, satiny-white hairs below. Flowers 2-4cm across occur in December and January. An abundant plant forming quite extensive patches in fellfields, grasslands and especially on dry exposed rocky slopes from Arthur's Pass to the Humboldt Mountains.

Musk daisies *Celmisia du-rietzii*
Celmisia discolor

Two closely related, strongly musk-scented, prostrate or sprawling, woody daisies. The lower sections of the branches bear leaf remains and the upper sections are crowded with rosulate tufts of imbricate, often sticky, leaves, 2-4cm long by 8-12mm wide, with their lower surfaces clothed by dense, satiny hairs. In *C. discolor* the upper leaf surface is lightly clothed with white hairs, but in *C. du-rietzii* this surface is normally smooth. Flowers 2-3cm across on stalks 10-15cm high occur during December and January. Found throughout the South Island mountains in alpine herbfields, fellfields, grasslands and rocky places, often forming huge patches.

373. A plant of Hector's daisy in full flower, Gertrude Cirque, 1,500m (January)

Hector's daisy   *Celmisia hecto*

A much-branching, low shrub forming patches up to 1m across. Branche are clothed with leaf remains; branchlets bear rosettes of leathery leave 15-20mm long by 5-7mm wide. Upper surfaces of leaves are clothed wit scale-like, appressed, white hairs, lower surfaces with satiny-white hair A single flower, 2-2.5cm across, arises from each rosette during Decembe and January. Found in alpine herbfields, rocky places and grasslands fro Arthur's Pass to Fiordland.

Boulder Lake daisy   *Celmisia par*

Found only in the vicinity of Boulder Lake and the Paparoa Mountain in boggy herbfields and grasslands, this tufted daisy often forms extensi patches. The branches are clothed with leaf remains; living leaves are 15cm long by 10-15mm wide, narrow-oblong to elliptic, tapering to ape with recurved, faintly toothed margins; the upper surface is smooth ar shining, the lower has dense, appressed, white hairs. Flowers 2-3cm acro occur in January and February.

Hybrid mountain daisy   *Celmisia compac*

Mountain daisies will readily hybridise, and the plant illustrated her known as *C. compacta*, is a hybrid formed possibly by the interbreedir of *C. sessiliflora* with another species. Its leaves are 30-50mm long 5mm wide, and the flower head 2-4cm across. It occurs occasionally herbfields from Mt Arthur to Arthur's Pass.

374. Boulder Lake daisy in flower, Douglas Ridge, Boulder Lake, 1,550m (February)

375. Hybrid mountain daisy, Mt Lucretia, Lewis Pass, 1,500m (January)

376. Fiordland mountain daisy in flower, Homer Saddle (January)

Fiordland mountain daisy   *Celmisia verbascifolia*

A large, tufted daisy with thick, leathery leaves, 15-25cm long by 2.5-5cm wide, smooth and shining above with the main veins distinct; the lower surface is covered by velvety hairs. Flowers, 2-2.5cm across on ribbed hairy stalks 30-40cm long, appear during January and February. An attractive-looking *Celmisia* found throughout the mountains of west Otago and Fiordland in herbfields and rocky places.

377. Young leaves of spine-leaved daisy, Old Man Range (January)

Mt Egmont daisy   *Celmisia major* var. *brevis*
A tufted herb with thick, leathery leaves, 10-15cm long by 5-10mm wide,
tapering from base to acute apex, with distinct midrib, recurving margins,
and satiny appressed hairs below. Flowers 2-3.5cm across occur from
December to February. Found on Mt Egmont in herbfields, scrub and
scoria up to 1,400m. A similar plant, var. *major,* larger in all its parts,
with leaves up to 40cm long, occurs down to sea level and on the cliffs
around Manukau Harbour, Piha, the Great Barrier Island and Wharariki Beach,
near Cape Farewell.

378. Mt Egmont daisies in flower, Mt Egmont
     (December)

Spine-leaved daisy   *Celmisia brevifolia*
A woody, branching, sprawling shrub forming large, loose clumps with
the branches densely clothed by leaf remains. The thick leaves, 10-15mm
long by 6-9mm wide, are arranged with the upper erect and the lower
spreading or deflexed; they are smooth on the upper surface, satiny-hairy
below, with the margins bearing widely separated small spines or teeth.
Flowers 2-3cm across on short sticky stalks 4-8cm high occur from January
to March. Found in herbfields and rocky places between 1,400 and 1,850m
in the mountains of Canterbury and Otago.

379. Marlborough daisy, Woods
Gorge (December)

380. Sticky-stalked daisy, Brown Cow
Pass, Boulder Lake (February)

Marlborough daisy *Celmisia monroi*

A tall daisy, up to 20cm high, which grows as single plants or as masses of branching stock. Rather similar to *C. semicordata,* with large flower heads up to 40cm across. The leaves, 7-15cm long and 5-20mm wide, are rigid with the upper surface grooved longitudinally and covered by a thin, silvery pellicle or skin, the lower surface covered with close, silvery-white hairs. Found in subalpine regions of Marlborough, especially on Mt Benmore and at the Woodside Gorge, in grasslands and rocky places.

Sticky-stalked daisy *Celmisia hieracifolia*

A simple, tufted daisy with leathery leaves, 8-12cm long by 1-2cm wide, having shallowly toothed margins, their upper surfaces green and their lower surfaces clothed with white. to buff-coloured satiny hairs. Flowers, 3-4cm across on sticky, glandular, hairy stalks 5-25cm high, occur during January and February.

Downy daisy *Celmisia glandulosa*

A creeping and rooting daisy, forming rosettes of leaves 15-25mm long by 10-15mm wide. The leaves are green on both surfaces but covered by many minute glands; the margins are toothed or serrate. Flowers 2cm across arise on slender stalks 7-12cm high during December and January. Found throughout New Zealand in subalpine herbfields, fellfields, rocky places or grasslands.

381. Downy daisy in flower, Mt Egmont (January)

382. White daisy flowers, Mt Ruapehu (January)

383. Leaves and developing flower buds of white daisy, Mt Ruapehu (December)

384. Dainty daisy in flower, Cass (November)

Dainty daisy   *Celmisia gracilenta*

A slender, tufted daisy with tough but flexible leaves, 10-15cm long by 2-4mm wide, each tapering from the base to an acute apex, with recurved margins rolled almost to the midrib and covering the appressed, satiny hairs of the lower surfaces; upper surfaces have a silvery pellicle. Flowers, 1-2cm across on slender stalks 25-40cm high clothed with satiny-white hairs, occur from November to February. Found throughout New Zealand in herbfields and grasslands up to 1,550m altitude. This was the first *Celmisia* discovered in New Zealand. It was found by Captain Cook's expedition in March 1770 on the hills near Queen Charlotte Sound.

White daisy   *Celmisia incana*

Sometimes called 'mountain musk', this stout, woody-stemmed, branching daisy forms large, silvery-white mats up to 1-1.5m wide in subalpine and alpine herbfields, fellfields, exposed rocky places and grasslands from the Coromandel Mountains to Otago. The thick, leathery leaves, 25-40mm long by 12-15mm wide, are densely clothed all over with appressed white hairs. The flowers, 2-3.5cm across, arise on short, slender stems during December and January.

385. Larch-leaf daisy in flower, Arthur's Pass (January)

386. Pigmy daisy with flowers, Brown Cow Pass to Boulder Lake (February)

Larch-leaf daisy   *Celmisia laricifolia*

A small, branching, mat-forming daisy with branches clothed by leaf remains and branchlets bearing dense tufts of narrow, pointed, silvery-green, larch-like leaves, 10-15mm long by 1-1.5mm wide. The apical point is often broken off and the leaf margin is rolled over almost to the midrib, covering the thin layer of white hairs on the lower surface. Flowers 1-2cm across occur on very slender stalks during January and February. Found in alpine fellfields, exposed rocky places and grasslands throughout the mountains of the South Island.

Purple-stalked daisy   *Celmisia petiolata*

A large, tufted herb with smooth leaves, 7-15cm long by 3-4cm wide, bright green, smooth and silky above, white or buff below with soft, velvety hairs; purple lateral veins and a flat, purple petiole are usually present and are characteristic. Flowers 5-7cm across occur on purplish hairy stalks 15-20cm high during December and January. Found in herbfields, fellfields and grasslands along the Alps and western mountains of the South Island from the Spencer Mountains southwards.

387. Purple-stalked daisy in flower, Arthur's Pass (December)

Pigmy daisy   *Celmisia lateralis*

A sprawling daisy with slender, branching stems up to 30cm long, found in herbfields, fellfields and exposed rocky places from the northwest Nelson Mountains to the Paparoa Ranges. The thick, hairless leaves are densely overlapping and 6-8mm long by 1-1.5mm wide. Flowers 1-2cm across occur on slender stalks up to 8cm long during January and February.

388. Dagger-leaf daisy in flower, Douglas Ridge, Boulder Lake (February)

Dagger-leaf daisy   *Celmisia petrie*
A stout, tufted daisy with rigid, thick, sharply pointed, dagger-shaped
leaves, 20-30cm long by 1-2cm wide, which are bright green and hairles
above, with two prominent longitudinal ridges, and white below, with
appressed satiny hairs. Flowers 3-4cm across are borne on stout, woolly
stems, 20-50cm high, from December to February. Found from the
mountains of Nelson and Marlborough southwards to Fiordland.

Sticky daisy   *Celmisia viscose*
A stout, low, shrubby daisy forming patches up to 1m across. The branche
are clothed in leaf remains, with the branchlets ascending with tufts o
very sticky leaves 6-8cm long by 6-9mm wide. Flowers 2-4cm across occu
abundantly on stout, hairy, sticky scapes, 15-30cm long, from Decembe
to February. Found mainly along the eastern side of the South Islan
mountains in herbfields, fellfields and grasslands above 1,100m. This dais
is reputed to be very sticky all over but in my experience, apart from
the flower stalk, it is barely sticky at all. Certainly it is not as stick
as *C. haastii*.

389. Sticky daisy in flower, Torlesse Range, 1,500m
(January)

390. Sticky daisy plant, Torlesse Range, 1,400m (January)

391. Brown mountain daisy in flower, Spencer Mountains near Lewis Pass, 1,600m (January)

392. White cushion daisy in flower, Temple Basin, Arthur's Pass (January)

Brown mountain daisy   *Celmisia traversii*
This large and strikingly beautiful daisy is found in alpine herbfields and
grasslands from northwest Nelson to the Lewis Pass. The leaves, 15-25cm
long by 4-5cm wide, have their undersurfaces and margins densely clothed
with rich brown-coloured, velvety hairs, and the leaf margins narrow to
a purple-coloured petiole 10-15cm long. Flowers 4-5cm across are borne
on stout stems, up to 30cm high, which are also densely clothed with
brown, velvety hairs. Flowers occur during December and January.

Cotton daisy, cotton plant, tikumu   *Celmisia spectabilis*
A stout, rosulate, tufted herb sometimes forming large patches. The leaves,
10-15cm long by 1-2.5cm wide, are very thick and leathery with the upper
surface smooth and shining, the lower surface densely clothed by soft,
matted, buff or white hairs. Flowers 3-5cm across occur abundantly on
white, woolly stalks 8-25cm high. Found in subalpine and alpine herbfields,
fellfields and grasslands from Mt Hikurangi southwards to North Otago.

393. Cotton daisy plant, Mt Holdsworth (February)

White cushion daisy   *Celmisia sessiliflora*
A freely branching, perennial herb, with its branchlets covered by closely
overlapping, thick, rather rigid leaves, 1-3cm long by 1-2mm wide. These
form tight rosettes, which together make a broad flat mat up to 1m across
and 5-10cm high. The stalkless flowers, 1-2cm across, are borne at the
tips of the rosettes of leaves during December and January. Found
throughout the mountains of the South Island, in herbfields, fellfields and
grasslands up to 1,700m.

394. Dusky Sound daisy, Stillwater Basin, Fiordland (December)

Dusky Sound daisy   *Celmisia holoserice*
First described from Dusky Sound, this large daisy is found throughou
Fiordland in rocky places and herbfields from coastal to subalpine region:
The leaves are 15-30cm long and 4-6cm wide, with a shining upper surfac
and a cladding of bright, silvery hairs on the lower surface, the apice
pointed. Flowers 5-7cm across occur during December and January.

Fiordland rock daisy   *Celmisia inaccess:*
A plant forming large mats up to 1m across. The vivid, lush, green leave
are 2-6cm long and 1-2cm wide, with distinct teeth around their margins
The flowers are up to 5cm across and occur during December and January
Found growing mostly on limestone or marble, rocky outcrops, it is i
conspicuous plant because of its vivid colour and lush leaves. It occur
in the Wapiti Lake to Barrier Peaks area between the Doon and Stillwate
Rivers, Fiordland.

395. Fiordland rock daisy above Wapiti Lake, Fiordland (December)

396. Leaves of Fiordland rock daisy above Wapiti Lake, Fiordland (December)

397. Plant of Allan's daisy coming into flower, Mt Lucretia, 1,500m (January)

Allan's daisy   *Celmisia allani*

A loosely branched daisy, with dense leaf remains below on the branches and rosulate living leaves at the tips. The leaves, which measure 3-4cm long by 1-1.5cm wide, are thin, flexible, and more or less densely clothed on both surfaces with soft, white hairs. Flowers 3-4cm across occur on hairy stalks during January and February. Found in herbfields and grasslands from the Nelson mountains to the Lewis Pass.

398. Haast's daisy in flower, Temple Basin (January)

399. Crag-loving daisy, from a specimen taken from Eyre Creek and grown by Mr J. Anderson at Albury (December)

Crag-loving daisy   *Celmisia philocremna*

A plant forming a cushion up to 70cm across with overlapping, thick, rather succulent-like leaves 1.5-2.5cm long and 5mm wide. Flower stalks very hairy and bearing flower heads up to 3cm across. This daisy is found only in the Eyre Mountains between 900 and 1,800m. It grows in crevices and on ledges in steep rocky bluffs.

Haast's daisy   *Celmisia haastii*

A strong-growing, branching daisy, forming patches up to 75cm across. The leathery leaves, 4-7cm long by 10-15mm wide, are smooth and green above with distinct longitudinal grooves and clothed below with appressed, satiny hairs. The whole plant is very sticky. Flowers, up to 4cm across on very hairy stalks, occur during December and January. Found in herbfields, fellfields, rocky places and grasslands throughout the South Island mountains.

400. Lesser onion-leaved orchid, Mt Holdsworth, 1,450m (February)

Lesser onion-leaved orchid   *Prasophyllum colenso*
The main stem, 5-35cm high, is sheathed in a leek-like leaf for about
three-quarters of its height, after which the stem and the leaf separate.
Flowers 5mm long occur from November to January and are faintly sweet-
scented. Found abundantly in subalpine herbfields and grasslands from
the Volcanic Plateau southwards.

Family: Orchidaceae

Broad-leaved thelymitra   *Thelymitra decor*
Plant up to 50cm high with the single-keeled, 10mm-wide leaf overtopping
the flowers in short-stemmed specimens. Flowers 8-18mm across occur
as one- to ten-flowered racemes during November and December. Found
in herbfields and grasslands throughout New Zealand up to 1,250m.

Family: Orchidacea

Short-leaved thelymitra   *Thelymitra venosa*
A slender plant, 10-50cm high, with the very narrow, thick leaf much
shorter than the stem. Flowers 2cm across occur singly or up to four
at a time from December to February. Found from Rotorua southwards
in herbfields and grasslands, particularly round damp seepages, up to
1,550m.

Family: Orchidaceae

401. Flower of broad-leaved thelymitra, *T. decora*, Volcanic Plateau (December)

402. Flower of short-leaved thelymitra, *T. venosa*, Douglas Ridge, Boulder Lake. Notice the prominent veins in the flower petals. 1,550m (February)

403. Flowers of long-leaved thelymitra, *T. longifolia*, Volcanic Plateau (November)

Long-leaved thelymitra   *Thelymitra longifolia*

Plant from 7-45cm high, with the single, fleshy, lanceolate leaf, 3-18mm wide, overtopping the flowers in short-stemmed specimens. Flowers 8-18mm across occur as two- to sixteen-flowered racemes during November and December. Found in herbfields and grasslands throughout New Zealand up to 1,250m. Flowers may be blue, pink, or white. White-flowered plants are known as variety *alba*.

Family: Orchidaceae

Lyall's orchid   *Caladenia lyallii*
The plant, 10-30cm high, is clothed with long, soft hairs, and has a single
leaf, 3-6mm wide, arising from the base of the stem but much shorter
than the stem. Flowers 1.5-2.5cm across occur from December to January.
They may be pink as well as white. Found throughout the South Island
in damp or boggy places in alpine herbfields and grasslands between 800
and 1,550m.

Family: Orchidaceae

404. Flowers of Lyall's orchid, Arthur's Pass (December)

405. Flowers of snow marguerite, Arthur's Pass (December)

406. Typical plant of snow marguerite in flower, Cleddau Cirque (January)

407. Hybrid between snow marguerite and yellow
snow marguerite, Homer Cirque (January)

Hybrid snow marguerites
Hybridisation occurs freely between *Dolichoglottis lyallii* and *D. scorzoneroides,* and all shades of intermediate forms can occur. In general, those with pale cream to creamy-white flowers are dominated by *D. scorzoneroides*, those with deeper-cream to yellowish flowers by *D. lyallii*. A plant, therefore, with *D. scorzoneroides* leaves and yellowish flowers is a hybrid; similarly a plant with *D. lyallii* leaves but pale cream or whitish flowers is also a hybrid. Types of leaves intermediate between the two species also occur in these hybrid plants.

Family: Compositae

Snow marguerite *Dolichoglottis scorzoneroides*
A leafy plant, reaching 60cm high, with stout, leafy stems, hairy in the upper part and terminated by broad corymbs of white flowers each 4-6cm across. The broad, fleshy, pointed leaves are up to 20cm long by 2-3cm wide (fig. 406). The flowers occur during December, January and February. Found in damp situations in subalpine and alpine herbfields, fellfields and grasslands all along the Southern Alps and in Stewart Island.

Family: Compositae

408. Hybrid between yellow snow marguerite and snow marguerite, Homer Cirque (January)

Yellow snow marguerite *Dolichoglottis lyallii*
Plant up to 50cm high with slender stems terminated by broad corymbs of brilliant, yellow flowers. The daffodil-like, fleshy leaves are up to 25cm long by 1cm wide. Each flower is 4-5cm across, and flowers appear from December to February. Found in the South Island in damp situations in herbfields, fellfields, rocky places and grasslands all along the Southern Alps and in Stewart Island.

Family: Compositae

409. Yellow snow marguerite in flower, Homer Saddle, 1,250m (January)

410. Flowers of yellow snow marguerite, Homer Cirque (January)

411. Flowers of yellow rock daisy,
Mt Holdsworth (December)

412. Plant of yellow rock daisy, Mt Holdsworth (January)

Yellow rock daisy   *Brachyglottis lagopus*
A herb clothed with silky hairs and easily recognised by its soft, reticulated,
leathery leaves, 3-15cm long by 3-10cm wide, on stout, hairy petioles 3-
10cm long. Flowers, 2-4cm across on hairy stems up to 35cm high, occur
from January to March. Found in alpine herbfields, fellfields, rocky places
and grasslands, mostly in the shade of other plants, from the Ruahine
Mountains to North Otago.

Family: Compositae

413. Flowers of dwarf alpine daisy, Arthur's Pass (January)

Dwarf alpine daisy   *Brachyglottis bellidioides*
A small alpine herb with the leaves appressed to the ground and a slender
flower stalk rising to 30cm high. The leaves are leathery, 1-5cm long,
with hairy margins and occasional hairs on the upper surface. The flowers,
2-3cm across, occur from October till March. Found throughout the South
Island in alpine herbfields, fellfields and grasslands; very common around
Arthur's Pass.

Family: Compositae

414. Plants of dwarf alpine daisy in flower, Arthur's
Pass (January)

415. New Zealand bluebell in flower, Arthur's Pass
(December)

416. New Zealand bluebells in tussock grassland, Arthur's Pass
(December)

417. Maori bluebell in flower, Volcanic Plateau
(November)

Maori bluebell   *Wahlenbergia pygmaea*
A low-growing, hairless, tufted, perennial herb, 1-2cm high, with linear-
spathulate, crenulate- or serrate-margined leaves, 15mm long including
petiole, 2-3mm wide, arranged as rosettes. Flowers 10-12mm long, held
erect on slender stalks 2-5cm high, occur from November to February.
Found in subalpine and alpine herbfields and grasslands from the Volcanic
Plateau south to Fiordland.
                                              Family: Campanulaceae

New Zealand bluebell   *Wahlenbergia albomarginata*
A small, tufted, perennial herb with elongate, elliptic or spathulate leaves
arranged as rosettes. Leaves 5-40mm long, including petiole, by 1-10mm
wide; the leaves become progressively smaller the drier the situation in
which the plant is growing. The leaf margins are slightly thickened and
bear 1-2 blunt teeth down each side. Flowers 1-3cm across may be erect,
inclined or drooping, depending upon the situation, and are borne on
slender stems 3-25cm high from November till February. Found throughout
the South Island mountains, from 600 to 1,550m, in herbfields, fellfields
and grasslands.
                                              Family: Campanulaceae

418. Brockie's bluebell, from a plant grown at Otari by Mr Brockie (January)

Brockie's bluebell   *Wahlenbergia brockie*
A perennial, tufted plant with underground interlacing branches producing
on the surface leafy rosettes, which may crowd together to form a clump.
Leaves elongate, 1-3cm long by 1-2mm wide. Flowers 16-20mm across
arise singly from each leafy rosette on slender stems 5-10cm high during
November and December. Found only on limestone soils near Castle Hill.
Family: Campanulacea

Large-flowered mat daisy   *Raoulia grandiflora*
A much-branched, creeping and rooting plant, forming mats or cushions
up to 15cm across, on rocks in herbfields, fellfields and exposed places
between 900 and 2,000m, from Mt Hikurangi southwards to Fiordland.
The densely overlapping leaves, 5-10mm long by 1-2mm wide, arranged
as rosettes, are clothed on both surfaces with silvery-white, appressed hairs.
Stalkless flowers, 8-16mm across, occur singly on the rosettes of leaves
during December and January.

Family: Compositae

Woollyhead   *Craspedia uniflora*
A perennial plant having soft leaves 5-12cm long, with their margins
whitened by tangled, woolly hairs. Woolly flower stalks, 15-45cm high,
bear rounded flower heads, 1.5-3cm in diameter, during December and
January. Found from the East Cape southwards, mostly in subalpine and
alpine herbfields, fellfields and exposed stony places up to 1,600m altitude.
Family: Compositae

419. Large-flowered mat daisy in flower, Mt Lucretia, Lewis Pass, 1,500m (January)

420. Woollyhead in flower, Gertrude Cirque, Homer (January)

421. Flowers of New Zealand chickweed, Cass (November)

New Zealand chickweed *Stellaria gracilent*
An erect, stiff, wiry, perennial herb, up to 10cm high, with thick, awl
shaped leaves, concave above and only 3-6mm long. Flowers 10-12mm
across arise singly on stems up to 5cm long from October to March. Found
in dry, rocky well-drained places in herbfields, fellfields and grassland
up to 1,550m from Mt Hikurangi to Fiordland.

Family: Caryophyllacea

Grassland daisy *Brachycome sinclairii* var. *sinclairi*
This bright little daisy differs from var. *pinnata* in the rosulate arrangemen
of the leaves. Flowers 6-15mm across occur on stalks 10-15mm long. Found
in herbfields and subalpine grasslands up to 1,700m throughout New
Zealand but relatively rare in the north. The var. *pinnata* is found only
in the South Island and Stewart Island.

Family: Compositae

422. Grassland daisy, var. *sinclairii*, Mt Lucretia, Lewis Pass, 1,600m (January) ▶

423. Plant of matted ourisia in flower, Mt Ruapehu
(December)

424. Flower of red-tinged variety of matted ourisia, Mt Ruapehu
(December)

425. Alpine fern, Homer Saddle, 1,300m (January)

Alpine fern *Polystichum cystostegia*

A fern with fronds 10-25cm long, which spreads by a branching, underground stem and which is often met with in rocky and stony places or amongst rocks in herbfields and fellfields in the South Island mountains between 900 and 1,700m.

Matted ourisia *Ourisia vulcanica*

A much-branched, spreading and rooting herb, forming a leafy mat 10-15cm across. The thick, fleshy, hairless leaves, 10-25mm long by 6-15mm wide, are appressed to the ground. Flowers, 1.5-2cm across on hairy stalks 5-12cm long, occur from October to January. Unlike other species of *Ourisia*, this plant grows in open sun in dry places and is found in herbfields, fellfields and exposed, rocky places on the Kaimanawa Mountains and the Volcanic Plateau up to 1,500m altitude. The flowers generally are white, but the plant shown in figure 424 is unusual and is the only one I have seen with this strong, red shading on the corolla.

Family: Scrophulariaceae

426. Flowers of mountain pinatoro, Sugarloaf, Cass (November)

Mountain pinatoro   *Pimelia oreophila*

The normal pinatoro plant is a prostrate, spreading shrub, forming mats
in lowland and subalpine regions. The species *oreophila*, common in
mountain regions, forms a small shrub 10-15cm high with ascending
branches. The older branches are scarred with rings left by falling leaves,
and the younger branches are usually hairy. The thick leaves, 3-6mm long
by 1-3mm wide, have silky-haired margins, often coloured with red when
fresh. The sweet-scented flowers, about 6mm across, which occur from
October to March, give rise to ovoid, white berries, about 2mm long,
from January onwards. Found throughout New Zealand in herbfields,
grasslands and exposed rocky places up to 1,600m altitude.

Family: Thymelaeaceae

427. Mountain pinatoro in flower, Sugarloaf, Cass (November)

PLANT ASSOCIATIONS

Most New Zealand plants will grow singly as isolated specimens but, in their natural state, they usually grow in intimate and often complicated associations.

428. Stunted beech forest, Key Summit (January). As higher altitudes are approached, the forest trees become increasingly more stunted and, in exposed alpine places of the South Island, beech trees many years old are often only 2-4m high. On such places as Key Summit the high precipitation enables mosses and lichens to clothe everything, including these gnarled and dwarfed old trees, with a soft, green mantle.

429. Subalpine scrub association, above Arthur's Pass (April). Apart from their immense botanical interest, dense plant associations such as these serve to shield and protect the alpine slopes from the elements, preventing erosion and keeping the waters of mountain streams and rivers clear and clean.

430. Subalpine scrub can be colourful, even when not in flower, as shown here clothing a gully on Mt Tongariro (September).

431. From scrub to tussock grassland — above Arthur's Pass. Alpine scrub merges into a tussock grassland with *Cassinia vauvilliersii* flowering in the foreground (January).

433. *Chionochloa pallens* and *Dracophyllum uniflorum* clothe a spur above the bush line, Mt Holdsworth (February).

434. A portion of the extensive alpine bogs on Key Summit, with large hummocks of sphagnum moss and containing *Celmisia alpina*, *Drosera* spp., and many other small, water-loving plants (January).

◀ 432. Streamside association of red tussock, mountain flax and *Olearia lacunosa*, near Boulder Lake (February). Such associations are typical of mountain streamsides in tussock country where the vegetation is healthy, unharmed by stock or introduced mammals. These plants form a giant 'sponge', which absorbs and regulates the flow of water into the stream during and after rain.

435. Alpine cushion, *Donatia novae-zelandiae*, in flower, and *Drosera stenopetala* in an alpine bog, Key Summit (January).

436. Mat of alpine cushion, *Donatia novae-zelandiae*, in alpine bog with common nertera (*Nertera balfouriana*) growing through it and bearing its ripe, red drupes, Arthur's Pass (April).

438. Only 15cm across but illustrating the intimacy and complexity of ▶ plant associations, *Phyllachne colensoi* is penetrated by plants of *Celmisia gracilenta* and *Raoulia grandiflora*. Mt Lucretia, Lewis Pass, 1,550m (January).

437. Wahu, *Drosera arcturi*, association in a
sphagnum bog, Lewis Pass (January).
(Refer also to fig. 239.)

439. An association of alpine plants growing along a crack carrying a water seepage through a glaciated rock face, Gertrude Cirque, 1,400m. *Aciphylla congesta*, *Dolichoglottis scorzoneroides*, and *Celmisia hectori* are the principal plants (January).

440. The dense, tight and lush clothing on the higher slopes of Mt Lucretia, Lewis Pass, provided by a healthy alpine meadow formed principally of *Celmisia discolor*, *Chionochloa australis*, and *Dracophyllum pronum* (January).

441. A typical herbfield association, Mt Lucretia, Lewis Pass, 1,500m, with *Dolichoglottis scorzoneroides* in flower, *Hebe lycopodioides*, and *Celmisia traversii* in the foreground.

442. Looking into the 'micro-ecology' of plants, we see buds of native pincushion, *Cotula squalida*, amongst *Gunnera dentata*, *Myriophyllum pedunculatum*, and other small, damp-rooting plants, Boulder Lake (February).

443. A giant vegetable sheep, *Haastia pulvinaris*, being invaded by tussocks, *Celmisia* and other plants as a scree becomes stabilised. This illustrates the inevitable changing successions of plant life in such places. Cupola Basin, 1,600m (April).

444. A pigmy speargrass, *Aciphylla polita*, associated with *Euphrasia townsonii*. Brown Cow Pass to Boulder Lake (February). *Euphrasia* plants are parasites on the roots of other plants, and *Euphrasia townsonii* here is living upon the *Aciphylla*.

445. Fellfield association of *Wahlenbergia albomarginata* with *Celmisia hectori*, Homer Saddle (January).

446. A New Zealand violet plant, *Viola cunninghamii*, growing and flowering in association with everlasting daisy, *Helichrysum bellidioides*, and the alpine tutu, *Coriaria plumosa*, Homer Cirque (December).

447. A giant buttercup, *Ranunculus lyallii*, in full flower in a fellfield on Mt Belle, Homer, beside plants of the horrid spaniard, *Aciphylla horrida*, and *Hebe odora*.

448. A colony of the yellow snow marguerite, *Dolichoglottis lyallii*, covering an alpine slope, Homer Saddle (January).

449. The intimacy and lushness of plant associations is well illustrated by this view of the creeping eyebright, *Euphrasia drucei*, growing on and over the North Island edelweiss, *Leucogenes leontipodium*, with *Celmisia spectabilis* and other small plants. Mt Holdsworth, 1,500m (December).

450. Scrub-herbfield association of large mountain daisy, *Celmisia semicordata*, growing luxuriantly and flowering amongst alpine scrub of *Hebe*, mountain flax, *Aristotelia fruticosa* and tussock. Arthur's Pass (January).

451. The musk daisy, *Celmisia discolor*, covers a low bank and makes a splendid display of flowers in an alpine herbfield. Mt Lucretia, Lewis Pass, 1,300m (January).

452. An alpine herbfield with *Celmisia semicordata* and *Bulbinella*, Jack's Pass (January).

453. A subalpine meadow with *Hebe odora*, snow marguerites, *Bulbinella* and other plants, Upper Hollyford Valley (January).

454. Creeping coprosma, *Coprosma pumila*, with red drupes, growing through a mat of white cushion daisy, *Celmisia sessiliflora*, on the Homer Saddle (January).

455. Plants of *Gingidium montanum* and mountain flax, *Phormium cookianum*, growing in close association at Arthur's Pass (December).

456. Completely covering the rocks, a lush association of *Celmisia verbascifolia, Dolichoglottis lyallii, Dracophyllum menziesii, Muehlenbeckia axillaris* and *Hymenanthera alpina* clothe a rocky substrate and make a huge absorbent sponge for water in the Homer Cirque (January).

457. *Celmisia hectori* cascades over a large boulder, enveloping a small plant of *Chianochloa conspicua* and with *Celmisia verbascifolia* in the foreground; a fellfield association from the Homer Saddle (January).

458. View north over the Lewis Pass showing typical sharp bush-demarcation line between the beech forest and the subalpine-alpine regions, with a tussock herbfield in the foreground (January).

459. Lichens are common plants that cover rocks like crusts in all alpine regions. They cling tightly to the rock surfaces and take on many forms and colours. They are important in changing rocks to soil, accomplishing this over very long periods of geological time. These lichens were growing on rocks in the Homer Cirque (January).

325

SELECTED REFERENCE WORKS

Allan, H.H., *Flora of New Zealand: Vol. 1.* New Zealand Government Printer, Wellington, 1961.

Brooker, S.G. and Cooper, R.C., *New Zealand Medicinal Plants.* A Handbook of the Auckland War Memorial Museum, Auckland, 1961.

Cheeseman, T.F., *Manual of the New Zealand Flora.* New Zealand Government Printer, Wellington, 1925.

Cockayne, L., *The Vegetation of New Zealand.* Third edition, Engelmann, Leipzig, 1958.

Cockayne, L. and Turner, E.P., *The Trees of New Zealand.* Second edition, New Zealand Government Printer, Wellington, 1958.

Hamlin, B., *Native Trees.* A.H. and A.W. Reed, Wellington, 1962.

Laing, R.M. and Blackwell, E.W., *Plants of New Zealand.* Fourth edition, Whitcombe and Tombs Ltd., Wellington, 1940.

Mark, A.F. and Adams, N. M., *New Zealand Alpine Plants.* A.H. & A.W. Reed, Wellington, 1973.

Moore, L.B. and Edgar, E. *Flora of New Zealand: Vol. 2.* New Zealand Government Printer, Wellington, 1970.

Moore, L.B. and Irwin, J.B., *The Oxford Book of New Zealand Plants.* Oxford University Press, Wellington, 1978.

Nordenstam, B., Taxonomic Studies in the Tribe Senecioneae (Compositae). *Opera Botanica,* Vol 44, pp 1-84, 1978.

Philipson, W.R. and Hearn, D., *Rock Garden Plants of the Southern Alps.* Caxton Press, Christchurch, 1962.

Poole, A.L. and Adams, N.M., *Trees and Shrubs of New Zealand.* Government Printer, Wellington, 1963. Revised 1979.

Richards, E.C., *Our New Zealand Trees and Flowers.* Third edition, Simpson and Williams Ltd., Christchurch, 1956.

Salmon, J.T., *New Zealand Flowers and Plants in Colour.* A.H. & A.W. Reed, Wellington, 1963.

Salmon, J.T., *The Native Trees of New Zealand.* A.H. & A.W. Reed, Wellington, 1980.

Wilson, H.D., *Field Guide, Wild Plants of Mount Cook National Park.* Field Guide Publication, Christchurch, 1978.

INDEX

North Cape

Waipoua kauri forest

Great Barrier I.

Waitakere Range
Auckland
Coromandel Range

Tauranga
Raukumara Rar
Mt Hikurangi
Kawhia
Opotiki
East Cap

Mokau R.
Lake Taupo
Mt Tauhara
Central Volcanic Plateau
Upper Waipunga Go
Mt Egmont
Tongariro National Park
Mahia Pen.
Mt Tongariro
Kaweka Mts
Mt Ruapehu
Kaimanawa Mts
Mt Ngauruhoe
Wanganui

Ruahine Range
Wharite Peak

Tararua Mts
Castlepoint
WAIRARAPA
Mt Holdsworth
Wellington
Mt Hector

0		200 km